Praise for
What a Fish Knows

One of the 10 Best Popular Science Books of 2016 in Biological Sciences, *Forbes*

One of the Week's Best Science Picks, *Nature*

"Latest Reads to Pique Your Curiosity," *Toronto Star*

"We Buddhists consider all animals, including fish, as sentient beings who have feelings of joy and pain just as we humans do. We also believe that they have all been kind to us as our mothers many times in the past, and are deserving of our compassion. Therefore, we try to help them in whatever way we can and at least avoid doing them harm. In *What a Fish Knows*, Jonathan Balcombe vividly shows that fish have feelings and deserve consideration and protection like other sentient beings. I hope reading it will help people become more aware of the benefits of vegetarianism and the need to treat animals with respect."

—The Dalai Lama

"[An] exhaustively researched and elegantly written argument for the moral claims of ichthyofauna."

—Nathan Heller, *The New Yorker*

"*What a Fish Knows* will leave you humbled, thrilled, and floored. Jonathan Balcombe delivers a revelation on every page, presenting jaw-dropping studies and stories that should reshape our understanding of, and compassion for, some of the most diverse and successful animals who have ever lived. After reading this, you will never be able to deny that fishes love their lives as we love ours, and that they, too, are vividly emotional, intelligent, and conscious. Bravo!"

—Sy Montgomery, author of *The Soul of an Octopus*, a National Book Award finalist

"Beautiful . . . We're much more similar to fish than meets the eye."

—David Gruber, Ideas.TED.com ("What Should You Read This Summer?")

"As ethologist Jonathan Balcombe notes in this engrossing study, breakthroughs are revealing sophisticated piscine behaviours. Balcombe glides from perception and cognition to tool use, pausing at marvels such as ocular migration in flounders and the capacity of the frillfin goby (*Bathygobius soporator*) to memorize the topography of the intertidal zone." —Barbara Kiser, *Nature*

"Balcombe covers the waterfront, so to speak, from fish cognition and perception to their social structures and breeding practices, all the while drawing on a dizzying array of experiments and studies. In the hands of a lesser writer, the sheer weight of material could have overburdened the reader. But Balcombe's prose is lively and clear, showcasing his gift for pithy sentences."

—Eugene Linden, *The American Scholar*

"*What a Fish Knows* bubbles with astounding fish facts."

—Kate Horowitz, *Mental Floss*

"*What a Fish Knows* seeks to acquaint us with the 'fabulous diversity' of sentient beings in our waters."

—Sarah Murdoch, *Toronto Star*

"The simple fact that fish live in an alien environment has created an information gap that scientists have been hard-pressed to bridge. Until now. Jonathan Balcombe, a professor of animal studies, fills the void in his new book *What a Fish Knows*, which argues we're not as different from our water-brethren as you'd think." —Joselin Linder, *New York Post*

"[An] eye-opening look at the lives of fish."
—Christopher Hart, *The Times* (London)

"*What a Fish Knows* . . . certainly left this piscivorous angler queasy about picking up his rod. There are other ways of interacting with these marvelous animals . . . Perhaps we should treat our aquatic kin with a bit more respect."
—Ben Goldfarb, *Hakai Magazine*

"With the vivacious energy of a cracking good storyteller, Balcombe draws deeply from scientific studies and his own experience with fish to introduce readers to them as sentient creatures that live full lives governed by cognition and perception . . . Balcombe makes a convincing case that fish possess minds and memories, are capable of planning and organizing, and cooperate with one another in webs of social relationships." —*Publishers Weekly*

"[A] sparkling exposition on 'our underwater cousins' . . . [and] a compelling pitch for greatly expanding fish conservation."
—Ray Olson, *Booklist*

"[Balcombe] offers an enjoyable, surprising and sometimes gruesome exploration of the world of fish, written with clarity and humor and grounded in many scientific studies . . . The breadth and depth of his research and his enthusiastic storytelling may permanently alter how [readers] look at a pet goldfish or a can of sardines." —Sara Catterall, *Shelf Awareness*

"I thought I knew a lot about fishes. Then I read *What a Fish Knows*. And *now* I know a lot about fishes! Stunning in the way it reveals so many astonishing things about the fishes who populate planet Earth in their trillions, this book is sure to 'deepen' your appreciation for our fin-bearing co-voyagers, the bright strangers whose world we share." —Carl Safina, author of *Beyond Words*

"Based on the latest scientific research, *What a Fish Knows* offers an eye-opening tour of the social, mental, and emotional lives of fishes. Who knew fishes use tools, appreciate music, fall for the same optical illusions we do, and engage in both cooperative hunting and some very kinky sex? Jonathan Balcombe's book is popular science writing at its best. It will spin your head around."

—Hal Herzog, author of *Some We Love, Some We Hate, Some We Eat*

"*What a Fish Knows* is a delightful and fascinating book that should be read by all who have dismissed fishes, especially the smaller denizens of the ocean, as utterly simple, primitive creatures. Jonathan Balcombe's lively descriptions of fish behavior are backed by solid science. What Carl Safina's *Beyond Words* did for elephants, wolves, and orcas, Balcombe's book does for fishes. It is a terrific read."

—Wendy Benchley, ocean conservationist and cofounder of the Peter Benchley Ocean Awards

"Outstanding. This excellent book brings fishes into their proper and well-deserved perspective." —Dave DeWitt, food historian

"Fishes are greatly misunderstood and grievously maligned. Now, in *What a Fish Knows*, Jonathan Balcombe uses the latest science to provide a comprehensive picture of just who fishes are. You will learn that fishes have distinct personalities, experience a wide range of emotions, form intricate social relationships, and are wonderful parents. Indeed, this forward-looking and long-overdue book is an integral part of reconnecting with the fascinating animals with whom we share our magnificent planet."

—Marc Bekoff, author of *The Emotional Lives of Animals* and *Rewilding Our Hearts*

"*What a Fish Knows* is the best book on fishes I have ever read. Brimming with engrossing anecdotes and humor, Jonathan Balcombe's inspiring treatise takes the reader on a fascinating and deeply moving journey into the lives of fishes. Balcombe's eloquent, persuasive, highly readable tour de force has a single, luminous message: Fishes deserve more respect, care, and protection."

—Chris Palmer, author of *Shooting in the Wild* and *Confessions of a Wildlife Filmmaker*

"Jonathan Balcombe's breathtaking book should instill a sense of humility and enormous wonder and awe at the rest of creation."

—David Suzuki, scientist, environmentalist, and broadcaster

AMIE CHOU / FOR THE HSUS

Jonathan Balcombe

What a Fish Knows

Jonathan Balcombe is the director of animal sentience at the Humane Society Institute for Science and Policy and the author of four books, including *Second Nature* and *Pleasurable Kingdom*. A popular commentator, he has appeared on *Fresh Air* with Terry Gross, the BBC, and the National Geographic Channel, and in several documentaries, and is a contributor of features and opinions to *The New York Times, The Washington Post, The Wall Street Journal, Nature,* and other publications. He lives in Florida. Find his community page "What a Fish Knows" on Facebook, follow him on Twitter at @Jonathanpb1959, and visit his website at jonathan-balcombe.com.

ALSO BY JONATHAN BALCOMBE

The Exultant Ark:
A Pictorial Tour of Animal Pleasure

Second Nature:
The Inner Lives of Animals

Pleasurable Kingdom:
Animals and the Nature of Feeling Good

The Use of Animals in Higher Education:
Problems, Alternatives, and Recommendations

WHAT A FISH KNOWS

WHAT A FISH KNOWS

THE INNER LIVES OF OUR UNDERWATER COUSINS

JONATHAN BALCOMBE

SCIENTIFIC AMERICAN / FARRAR, STRAUS AND GIROUX

NEW YORK

Scientific American / Farrar, Straus and Giroux
18 West 18th Street, New York 10011

Printed in the United States of America
Published in 2016 by Scientific American / Farrar, Straus and Giroux
First paperback edition, 2017

An excerpt from *What a Fish Knows* originally appeared, in
slightly different form, in *Scientific American*.

The Library of Congress has cataloged the hardcover edition as follows:
Names: Balcombe, Jonathan P.
Title: What a fish knows : the inner lives of our underwater cousins /
Jonathan Balcombe.
Description: First edition. | New York : Scientific American/Farrar,
Straus, and Giroux, 2016. | Includes bibliographical references
and index.
Identifiers: LCCN 2015048629 | ISBN 9780374288211 (hardcover) |
ISBN 9780374714338 (e-book)
Subjects: LCSH: Fishes—Behavior. | Fishes—Physiology. | Animal
communication. | Animal intelligence.
Classification: LCC QL639.3 .B35 2016 | DDC 597.15—dc23
LC record available at https://lccn.loc.gov/2015048629

Paperback ISBN: 978-0-374-53709-8

Designed by Jonathan D. Lippincott

9 10

To the anonymous trillions

Contents

WHAT A FISH KNOWS

Prologue

When I was eight I climbed into an aluminum rowboat with the elderly director of a summer camp north of Toronto. He rowed a quarter mile out into the shallow bay, and we spent the next two hours fishing. It was a calm summer evening and the water was like glass. It was my first time in a small boat, and floating on this vast, faintly undulating expanse of dark water was exhilarating. I wondered what creatures lurked below, and this stoked my excitement whenever the sudden jerk of my primitive fishing pole—a stripped sapling with a line and hook—signaled that a fish had struck the bait.

I caught sixteen fish that day. Some we released. Several others, larger bass and perch, we kept for breakfast the next morning. Mr. Nelson did all the dirty work, baiting the barbed hooks with writhing earthworms, twisting the wire out of the fish's lips, plunging his knife into their skulls to kill them. His face contorted strangely as he performed these tasks, and I wondered if he was feeling revulsion or if he was merely lost in concentration.

I have some fond memories of that experience. But, as a sensitive boy with a soft spot for animals, I was disturbed by a lot of what went on in that rowboat. I worried privately about the worms. I fretted that the fish felt pain as the stubborn hook was extracted from their bony, staring faces. Maybe one of the "keepers" survived the knife and was dying slowly in the wire basket dangling over the side. But the kind man sitting at the bow didn't seem to think there was anything wrong, so I rationalized that it must be okay. And the taste of fresh fish at breakfast the next morning left only vague remnants of misgivings from the previous evening.

That was not my only childhood encounter with fish that raised conflicting emotions about our cold-blooded cousins' place in our moral calculus. In fourth grade, I was one of a few kids recruited to move some supplies from our classroom to a neighboring room at my elementary school in Toronto. Among the items was a glass fishbowl containing a lone goldfish. The vessel was three-quarters full of water, and quite heavy. Concerned that the fish not be placed in the hands of someone who might care less than I did, I volunteered to transport the bowl to its destination, a counter next to the sink in the adjoining room.

How ironic.

I firmly held the bowl in my child's hands and methodically walked out the door, down the hall, and into the new room. As I gingerly approached the counter, the bowl slipped from my grasp and smashed on the hard floor. It was a moment of horror that played out in slow motion. Shards of glass splintered and water sloshed across the floor. I stood there stunned. Someone with more wits than I grabbed a mop and moved the glass and water to one side, then four of us began to scour the floor for the fish. A minute went by with no sign of the creature. It was like a bad dream. It seemed as if she had experienced goldfish rapture and risen up to the fishy heavens. Finally, someone found her. She had bounced behind a radiator and ended up on the inside lip, two inches above the floor and completely out of view. She was still alive, gawping meekly. She was quickly plopped into a beaker of tap water. I believe that fish survived.

Though the goldfish incident left a deep impression on me, as evidenced by my vivid recollection of it four decades later, I was not moved to a new empathy for fish. Admittedly, I never took a shine to fishing; what little enthusiasm remained after the outing with Mr. Nelson soon faded when it came time to bait and extract my own hooks. But I made no connection between the perch and bass I unceremoniously hauled up from Sturgeon Bay, or the hapless little goldfish I dropped at Edithvale Elementary School, and the anonymous fish who ended up in the Filet-O-Fish sandwiches I enjoyed on family trips to the local McDonald's. That was the late sixties, and already McDonald's was boasting "over one billion served." They could as soon have been referring to fish or chickens as to customers. But like other members of my culture, I was blissfully removed from the once living, breathing creatures who ended up in my lunch.

It was not until I took an ichthyology course in the final year of my undergraduate biology degree twelve years later that I began to seriously question my relationship to animals, including fish. I was as captivated by fishes' diverse anatomy and adaptations as I was disturbed by the parade of inert, once-living bodies we were given to classify using dissecting microscopes and taxonomic keys. The class made a midterm visit to the Royal Ontario Museum, where we met one of Canada's foremost ichthyologists for a private tour of the museum's fish collection. At one point he unlocked and raised the lid of a large wooden case to reveal an enormous lake trout floating in an oily preservative. The fish, weighing a record 103 pounds, had been caught on Lake Athabasca in 1962. Her size and plumpness were attributed to a hormone imbalance that had rendered her sterile; energy that would normally have been spent on the profligate task of egg production was instead assigned to body mass.

I felt for that fish. Like most we encounter, she had no name and her life was a mystery. I felt like she deserved a more dignified existence than entombment in a wooden case. To me it would have been better had she been eaten, her tissues recycling back into the food chain, than to float for decades in darkness, polluted by chemicals.

Legions of books have been written about fish—their diversity, their ecology, their fecundity, their survival strategies. And several bookshelves can be filled with books and magazines about how to catch fish. To date, however, no book has been written *on behalf of* fish. I'm not referring to the conservationist message that decries the plight of endangered species or the overexploitation of fish stocks (have you ever noticed that the word "overexploitation" legitimizes exploitation, and that "stocks" reduces an animal to a commodity like wheat whose sole purpose is to supply humans?). My book aims to give voice to fish in a way that hasn't been possible in the past. Thanks to breakthroughs in ethology, sociobiology, neurobiology, and ecology, we can now better understand what the world looks like to fish, how they perceive, feel, and experience the world.

In researching this book I have sought to sprinkle the science with stories of people's encounters with fishes, and I will be sharing some of these as we go along. Anecdotes carry little credibility with scientists, but they provide insight into what animals may be capable of that science has yet to explore, and they can inspire deeper reflection on the human-animal relationship.

What this book explores is a simple possibility with a profound implication. The simple possibility is that fishes* are individual beings whose lives have intrinsic value—that is, value to themselves quite apart from any utilitarian value they might have to us, for example as a source of profit, or of entertainment. The profound implication is that this would qualify them for inclusion in our circle of moral concern.

Why bother? There are two main reasons. First, fishes are, collectively, the most exploited (and overexploited) category of vertebrate animals on Earth. Second, the science of fish sentience

* We traditionally refer to anything from two to a trillion fish by the singular term "fish," which lumps them together like rows of corn. I have come to favor the plural "fishes," in recognition of the fact that these animals are individuals with personalities and relationships.

and cognition has advanced to a point that it may be time for a paradigm shift in how we think about and treat fishes.

Just how exploited are they? One author, Alison Mood, has estimated, based on analysis of Food and Agriculture Organization fisheries capture statistics for the period 1999–2007, that the number of fishes killed each year by humans is between 1 and 2.7 trillion.* To get a handle on the magnitude of a trillion fishes, if the average length of each caught fish is that of a dollar bill (six inches) and we lined them up end to end, they would stretch to the sun and back—a round-trip of 186 million miles—with a couple hundred billion fishes to spare.

Mood's estimate is exceptional because the human toll on fishes is rarely presented as a number of individuals. To wit, the Food and Agriculture Organization itself estimated the 2011 commercial fisheries catch to be 100 million tons. Fish biologists Steven Cooke and Ian Cowx, among the few to enumerate individual deaths, estimated in 2004 that some 47 billion fishes were being landed *recreationally* worldwide every year, of which some 36 percent (about 17 billion) were killed and the remainder returned to the water. If we apply their estimated average weight per fish (0.635 kilograms = 1.4 pounds) to a commercial catch of 100 million tons, we arrive at an estimate of 157 billion individual fishes.

One study reports that official (FAO) statistics on global fish catches have been underestimated by more than half over the last sixty years, due to often-neglected small-scale fisheries, illegal and other problematic fisheries, and discarded bycatch.

However you slice it, it's a lot of fishes, and they do not die nicely. The leading causes of death for commercially caught fishes are asphyxiation by removal from the water, decompression from

*Mood's estimate does not include recreational fishing, fishes caught illegally, fishes caught as bycatch and discarded, fishes who die following escape from nets, "ghost fishing" by lost or discarded gear, fishes caught for the fisher's own use as bait but not recorded, and fishes caught (but not recorded) for use as feed on fish and shrimp farms.

the pressure change of being brought to the surface, crushing beneath the weight of thousands of others hoisted aboard in massive nets, and evisceration once landed.

Regardless of which estimate you take, dizzying numbers like these tend to mask the fact that each fish is a unique individual, not just with a biology, but with a biography. Just as each sunfish, whale shark, manta ray, and leopard grouper has a distinctive pattern from which you can recognize individuals on the outside, each has a one-of-a-kind life on the inside, too. And therein lies the locus of change in human-fish relations. It is a fact of biology that every fish, like the proverbial grain of sand, is one of a kind. But unlike grains of sand, fishes are living beings. This is no trivial distinction. When we come to understand fishes as conscious individuals, we may cultivate a new relationship to them. In the immortal words of an unknown poet: "Nothing has changed except my attitude—so everything has changed."

PART I

THE MISUNDERSTOOD FISH

We shall not cease from exploration
And the end of all our exploring
Will be to arrive where we started
And know the place for the first time.

—T. S. Eliot

What we casually refer to as "fish" is in fact a collection of animals of fabulous diversity. According to FishBase—the largest and most often consulted online database on fishes—33,249 species, in 564 families and 64 orders, had been described as of January 2016. That's more than the combined total of all mammals, birds, reptiles, and amphibians. When we refer to "fish" we are referring to 60 percent of all the known species on Earth with backbones.

Almost all modern fishes are members of one of two major groups: bony fishes and cartilaginous fishes. Bony fishes, scientifically termed *teleosts* (from the Greek *teleios*=complete, and *osteon*=bone), make up the great majority of fishes today, numbering about 31,800 species, including such familiar ones as salmons, herrings, basses, tunas, eels, flounders, goldfishes, carps, pikes, and minnows. Cartilaginous fishes, or *chondrichthyans* (*chondr* =cartilage, and *ichthys*=fish), number about 1,300 species, including sharks, rays, skates, and chimaeras.* Members of both groups have all ten body systems of the land-dwelling vertebrates: skeletal,

* Some scientists place the chimaeras, also known as ghost sharks, in a separate group.

muscular, nervous, cardiovascular, respiratory, sensory, digestive, reproductive, endocrine, and excretory. A third distinct group of fishes is the jawless fishes, or *agnathans* (*a* = without, and *gnatha* = jaws), a small division of about 115 species comprising lampreys and hagfishes.

We conveniently classify animals with backbones into five groups: fishes, amphibians, reptiles, birds, and mammals. This is misleading because it fails to represent the profound distinctions among fishes. The bony fishes are at least as evolutionarily distinct from the cartilaginous fishes as mammals are from birds. A tuna is actually more closely related to a human than to a shark, and the coelacanth—a "living fossil" first discovered in 1937—sprouted closer to us than to a tuna on the tree of life. So there are at least *six* major vertebrate groups if one counts the cartilaginous fishes.

The illusion of relatedness among all fishes is partly attributable to the constraints of evolving to move efficiently in water. The density of water is about 800 times greater than that of air, so aquatic living has, in vertebrates, tended to favor streamlined shapes, muscular bodies, and flattened appendages (fins) that generate forward propulsion while minimizing drag.

Living in a denser medium also greatly reduces the pull of gravity. The buoyant effect of water frees aquatic organisms from the ravages of weight on terrestrial creatures. Thus, the largest animals—the whales—live in water, not on land. These factors also help explain the small relative brain size (the ratio of brain weight to body weight) of most fishes, which has been used against them in our cerebrocentric view of other life forms. Fishes benefit from having large, powerful muscles to propel them through water, which is more resistant than air, and living in a practically weightless environment means there is no premium on limiting body size relative to brain size.

In any event, brain size is only marginally meaningful in terms of cognitive advancement. As the author Sy Montgomery notes in an essay on octopus minds, it is well known in electronics that anything can be miniaturized. A small squid can learn mazes faster than dogs do, and a small goby fish can memorize in one trial the topography of a tide pool by swimming over it at high tide—a feat few if any humans could achieve.

The earliest fishlike creatures arose in the Cambrian period, some 530 million years ago.* They were small and not very exciting. The big breakthrough in the evolution of fishes (and all their descendants) was the appearance of jaws about 90 million years later in the Silurian period. Jaws allowed these pioneer vertebrates to grab and break up food items and to expand their heads to powerfully suck in prey, which greatly extended the available dinner menu. We might also think of jaws as nature's first Swiss Army knife, for they come with other functions, including manipulating objects, digging holes, carrying material to build nests, transporting and protecting young, transmitting sounds, and communicating (as in, don't come any closer or I'll bite you). Having jaws set the stage for an explosion of piscine life during the Devonian period—also known as "the age of fishes"—including the first super-predators. Most of the Devonian fishes were *placoderms* (plate-skinned), having heavy, bony armor over the head end and a cartilaginous skeleton. The largest placoderms were formidable. Some species of *Dunkleosteus* and *Titanichthys* measured well over thirty feet. They had no teeth, but could shear and crush with two pairs of sharp bony plates forming the jaws. Their fossils are often found with boluses of semi-digested fish bones, suggesting that they regurgitated these in the manner of modern owls.

* It was another 100 million years before an intrepid lobe-finned descendant took its first tentative steps on land. To get a perspective on these time spans, consider that the genus *Homo* to which modern humans belong has been around for only about 2 million years. If we compress our time on Earth down to one second, fishes have been around for over four minutes. They had graced planet Earth fifty times longer than we have before they even left the water.

Although they all went out with the Devonian and have been gone for over 300 million years, nature was kind to the placoderms in preserving some specimens so delicately that paleontologists have been able to deduce some intriguing facets of their lives. One particularly revealing find, from the Gogo fossil sites of Western Australia, is *Materpiscis attenboroughi* (translation: Attenborough's mother fish), named for the iconic British nature documentary presenter David Attenborough, who waxed enthusiastic over this species in his 1979 documentary series *Life on Earth*. This perfectly preserved 3-D specimen allows careful peeling away of layers to reveal the insides of the fish. And what should show up there but a well-developed baby *Materpiscis attenboroughi* attached to its mother by an umbilical cord. This discovery rocked the evolutionary boat by setting back the origins of internal fertilization by 200 million years. It also eroticized the lives of early fishes. As far as we know there is only one way to achieve internal fertilization: sex with an intromittent organ. So it appears that fishes were the first to enjoy "the fun kind" of sex. About this discovery and John Long, the Australian paleontologist who brought it to light, Attenborough expressed ambivalence during a public lecture: "This is the first known example of any vertebrate copulating in the history of life . . . and he names it after me."

Sex notwithstanding, the bony fishes, which arose about the same time as the placoderms, had a brighter future. Although they suffered major losses during the third great extinction that closed out the Permian period, they steadily diversified over the next 150 million years of the Triassic, Jurassic, and Cretaceous periods. Then, about 100 million years ago, they truly began to flourish. From that time to today the number of known families of bony fishes has more than quintupled. Fossil records do not divulge their secrets willingly, however, so there may be many earlier fish families still hidden in the rocks.

Like their bony counterparts, the cartilaginous fishes also

steadily recovered from the Permian setback, albeit without the explosive diversification of later times. As far as we know, there are more kinds of sharks and rays today than at any point in history. And we're beginning to discover that their real lives belie their pugnacious reputation.

Diverse and Versatile

Because their lives are more difficult to observe than those of most terrestrial animals, fishes are not easily fathomed. According to the National Oceanic and Atmospheric Administration, less than 5 percent of the world's oceans have been explored. The deep sea is the largest habitat on Earth, and most of the animals on this planet live there. A seven-month survey using echo soundings of the mesopelagic zone (between 100 and 1,000 meters—330 to 3,300 feet—below the ocean surface), published in early 2014, concluded that there are between ten and thirty times more fishes living there than was previously thought.

And why not? You might have encountered the popular notion that living at great depths is a terrible hardship for the creatures there. It's a shallow idea, for surely deep-sea creatures are no more inconvenienced by the enormous pressure of the overlying ocean than we are by the approximately ten-tons-per-square-meter pressure (often expressed as 14.7 pounds per square inch) of the atmosphere above us. As the ocean ecologist Tony Koslow explains in his book *The Silent Deep*, water is relatively incompressible, so deep-sea pressures have less impact than we usually think, because pressure from within the organism is about the same as that on the outside.

Technology is just beginning to afford us a glimpse of the ocean depths, but even in reachable habitats many species remain undiscovered. Between 1997 and 2007, 279 new species of fishes were found in Asia's Mekong River basin alone. The year 2011 saw the discovery of four shark species. Given the current rate, experts predict the total count of all fishes will level off at around 35,000. With the advance of techniques for distinguishing species at the genetic level, I think it could be many thousands more than that. When I studied bats as a graduate student in the late 1980s, 800 species had been identified. Today, the count has ballooned to 1,300.

From diversity springs variety, and from the rich variety of fishdom spring some noteworthy superlatives and bizarre life-history patterns. The smallest fish—indeed, the smallest vertebrate—is a tiny goby of one of the Philippine lakes of Luzon. Adult *Pandaka pygmaea* are only a third of an inch in length and weigh about 0.00015 of an ounce. If you were to put 300 of them on a scale they wouldn't equal the weight of an American penny.

At less than half an inch, some male deep-sea anglerfishes are not much bigger, but what they lack in size they make up for in the sheer audacity of their mode of existence. On finding a female, males of some deep-sea anglerfish species latch their mouths onto her body and stay there for the remainder of their lives. It doesn't matter much where they fix their bite on the female—it could be on her abdomen or her head—they eventually become fused to her. Many times smaller, the male resembles little more than a modified fin, living off her blood supply and fertilizing her intravenously. One female may end up with three or more males sprouting from her body like vestigial limbs.

It looks like a lurid form of sexual harassment; scientists have called it *sexual parasitism*. But the origins of this unconventional mating system are not so ignoble. It is estimated that female deep-sea anglerfishes occur at a density of about one per 800,000 cubic meters (28 million cubic feet) of water, which means a male is searching for a football-size object in a darkened space about the volume of a football stadium. Thus, it is desperately hard for anglerfishes to find each other in the vast darkness of the abyss, making it wise to hang on to your partner if you find one. At the time that Peter Greenwood and J. R. Norman revised *A History of*

Fishes in 1975, no free-swimming adult male anglerfish had been found, leading ichthyologists to speculate that the only alternative to successful latching is death. But the University of Washington's Ted Pietsch—curator of fishes at the Burke Museum of Natural History and Culture, and the world's leading authority on deep-sea anglerfishes—tells me that there are now hundreds of (formerly) free-living males in specimen collections around the world.

In exchange for the male being the ultimate couch potato, the female never has to wonder where her mate is on a Saturday evening. It turns out that some males do indeed amount to little more than an appendage.

Another fish superlative is their fecundity, which is also unmatched among vertebrates. A single ling, five feet long and weighing fifty-four pounds, had 28,361,000 eggs in her ovaries. Even that pales compared to the 300 million eggs carried by an ocean sunfish, the largest of all bony fishes. That such a grand creature can be the product of such a paltry parental investment as a teeny egg released into the water column might contribute to the common bias that fishes are unworthy of our consideration. But it bears reminding that all living things start from a single cell. And as we'll see in the section on "Parenting Styles," parental care is well developed in many fishes.

From its humble beginnings as an egg smaller than this letter "o," a mature adult ling can grow to be close to six feet long, and it is another superlative of fishes that they can increase so much in size from the start of their independent life cycle. But the growth champion among vertebrates may be the pointed-tailed ocean sunfish. While not streamlined (the family name, Molidae, refers to their millstone shape), they grow from one-tenth of an inch to ten feet in length, and can weigh 60 million times more as an adult.

Sharks lie at the opposite end of the spectrum of fish fecundity. Some species reproduce at a rate of only one baby a year. And that's only after they reach sexual maturity, which for some species can take a quarter century or more. In parts of their range, spiny dogfish sharks—a heavily fished species that you might have dissected in a college biology course—average thirty-five years old before they are ready to breed. Sharks have a placental structure as complex as that of mammals. Pregnancies are few and far between, and gestation can be lengthy. Frilled sharks carry their babies for over three years, the longest known pregnancy in nature. I sure hope they don't get morning sickness.

Dogfishes can't fly, nor can any other fishes, but they just might be the world's superlative for gliding. Best known of these are the flying fishes, of which there are about seventy species inhabiting the surfaces of the open ocean. Flying fishes have greatly enlarged pectoral fins that function as wings. In preparation for launch, they can reach speeds of forty miles per hour. Once airborne, the lower lobe of the tail may be dipped into the water and used as a supercharger to extend flights to 1,200 feet or more. Flights are usually just above the surface, but sometimes gusts of wind carry these aerialists fifteen to twenty feet high, which may explain why they sometimes land on ship decks. I wonder if the respiratory limitations of being a water breather have kept flying fishes from becoming truly flapping their "wings" for fully sustained flight? Fishes of several other types also launch themselves into the air, including the characins of South America and Africa, and—never mind that their name sounds more like a circus act—the flying gurnards.

Speaking of superlatives, and names, surely one of the longest belongs to Hawaii's state fish, the rectangular triggerfish, known by the locals as humuhumunukunukuapua'a (translation: the fish that sews with a needle and grunts like a pig). Perhaps the award for least flattering name should go to an anglerfish dubbed the hairy-jawed sack-mouth, and for most preposterous to the sarcastic fringehead. For the title of crudest, I nominate a small coastal dweller, the slippery dick (*Halichoeres bivittatus*).

But really, the most exciting breaking news on fishes is the steady stream of discoveries on how they think, feel, and live their lives. Scarcely a week now passes without a revealing new discovery of fish biology and behavior. Careful observations on reefs are uncovering nuanced social dynamics of cleaner–client fish mutualisms that defy the human conceit that fishes are dim-witted pea brains and slaves to instinct. And the notorious three-second fish memory has been debunked by simple laboratory investigations. In the pages ahead we'll explore how fishes are not just sentient, but aware, communicative, social, tool-using, virtuous, even Machiavellian.

Lowly Not

Among the vertebrate animals—mammals, birds, reptiles, amphibians, and fishes—it is the fishes that are the most alien to our sensibilities. Lacking detectable facial expressions and appearing mute, fishes are more easily dismissed than our fellow air breathers. Their place in human culture falls almost universally into two entwined contexts: (1) something to be caught, and (2) something to be eaten. Hooking and yanking them from the water has not just been seen as benign but as a symbol of all that's good about life. Fishing appears gratuitously in advertising, and the logo of one of America's most beloved film production studios, DreamWorks, features a Tom Sawyer-esque boy relaxing with a fishing pole. You may have met self-professed vegetarians who nonetheless eat fishes, as if there were no moral distinction between a cod and a cucumber.

Why have we tended to relegate fishes beyond the outer orbit of our circle of moral concern? For one thing, they are "cold-blooded," a layman's term that has little credibility in science. I do not see why having a built-in thermostat or not should have anything to do with an organism's moral status. In any event, most fishes' blood does not run cold. Fishes are ectothermic, meaning that their body temperatures are governed by outside factors, notably the water they are living in. If they live in warm tropical waters, their blood runs warm; if they live in the frigid reaches of the ocean depths or the polar regions, as many fishes do, then their body temperatures hover around freezing.

But even that description falls short. Tunas, swordfishes, and some sharks are partly endothermic—they can maintain body temperatures warmer than their surroundings. They achieve this by capturing heat generated by their powerful active swimming muscles. Bluefin tunas keep their muscle temperatures between 82 and 91 degrees Fahrenheit in waters ranging from 45 to 81 degrees. Similarly, many sharks have a large vein that warms the central nervous system by draining warm blood from the core swimming muscles to the spinal cord. The large, predatory billfishes (marlins, swordfishes, sailfishes, spearfishes) use this heat to warm their brains and eyes for optimal functioning in deeper, cooler waters. In March 2015, scientists described the first truly endothermic fish, the opah, which maintains its body temperature at about 9 degrees Fahrenheit above the cold waters it swims in at depths of several hundred feet, thanks to heat generated by the flapping of its long pectoral fins and conserved by a countercurrent heat exchange system in its gills.

Another prejudice we hold against fishes is that they are "primitive," which in this context has a host of unflattering connotations: simple, undeveloped, dim, inflexible, and unfeeling. Fishes were "born in front of my sunrise," wrote D. H. Lawrence in his 1921 poem "Fish."

No one is questioning that fishes have been around a long time, but therein lies the fallacy of labeling fishes as primitive. This bias presumes that those that stayed in the water stopped evolving the moment a few of them went ashore, a notion completely at odds with the tireless process of evolution. The brains and bodies of all extant vertebrates are a mosaic of primitive and advanced characteristics. Given time, and there's been plenty, natural selection keeps what works and winnows the rest, mainly through a process of gradual refinement.

All of the fish species that were living at the dawn of legs and lungs are long gone. About half of the fishes we see on the planet today belong to a group called the Percomorpha, which underwent an orgy of speciation just 50 million years ago (mya) and reached peak diversity around 15 mya, when the ape family, Hominoidea, to which we belong, was also evolving.

So about half of fish species are no more "primitive" than we are. But the descendants of early fishes have been evolving eons longer than their terrestrial counterparts, and on these terms fishes are the most highly evolved of all vertebrates. You might be surprised to know that fishes have the genetic machinery to make fingers—something that shows how similar fishes are to modern mammals. They just don't develop fingers, but fins instead, since fins are better for swimming than fingers are. And don't forget your segmented musculature. The *rectus abdominus*—the washboard stomach that graces the torso of our fittest athletes (and exists in all of us, albeit buried under a bit too much adipose tissue)—harkens back to the axial muscle segmentation first laid down by the fishes. As the title of Neil Shubin's popular book *Your Inner Fish* reminds us, our ancestors (and those of modern fishes) were early fishes, and our bodies are packed with modified structures traceable to those of our common aquatic forebears.

An older organism isn't necessarily simpler. Evolution does not trend relentlessly toward increased sophistication and size. Not only were the largest dinosaurs much larger than modern reptiles, paleontologists have recently unearthed evidence that they were social creatures with parental care and modes of communication at least as complex as those of modern reptiles. Similarly, the largest terrestrial mammals died out thousands or millions of years ago, at a time when mammalian diversity flourished. The true *age of mammals* is over. We tend to think of the last 65 million years as the Age of Mammals, but teleost fishes have been diversifying much more during that time. The Age of Teleosts may not sound quite as sexy, but it's more accurate.

Just as evolution does not proceed inescapably toward increased complexity, nor is it a process of perfection. For all the elegance with which adaptations allow animals to function optimally, it is a fallacy that animals are perfectly tailored to their environments. They can't be, because environments aren't static. Weather patterns, geological shifts such as earthquakes and volcanoes, and the constant process of erosion present moving targets. Even beyond these instabilities, nature is not fully efficient. There are inevitably compromises. Human examples include our appendix, our wisdom teeth, and the blind spot where the optic nerve interrupts the retina. For fishes, there is the closing of the gill covers necessary for respiration, which causes a forward thrust. If a fish wishes to remain stationary, as a resting fish usually does, she needs to compensate for the gill thrust. This is why you will rarely see a stationary fish whose pectoral fins are not in motion.

As we learn more about fishes, be it their evolution or their behavior, our capacity to identify with them grows, along with our ability to relate their existence to our own. Central to empathy—the capacity to place oneself in another's shoes, or in this case, fins—is an understanding of the experiences of the other. Central to that is an appreciation of their sensory worlds.

PART II

WHAT A FISH PERCEIVES

There is no truth. There is only perception. —Gustave Flaubert

What a Fish Sees

. . . red-gold, water-precious, mirror-flat bright eye
—from "Fish," by D. H. Lawrence

We are taught that there are five senses: vision, smell, hearing, touch, and taste. In truth, this is a restricted list. Think how dull life would be if you did not have a sense of pleasure! And while the idea of life without pain is appealing, how dangerous would it be if you didn't realize you were resting your hand on a burning hot stove? Without a sense of balance, we'd have no success walking, let alone bicycling. Without the ability to detect pressure, handling a knife and fork adeptly would become tasks requiring herculean feats of concentration. As we might expect for creatures who have had a lot of time to evolve, fishes have diverse, advanced modes of sensory perception.

One of my favorite concepts learned as a student of animal behavior is *umwelt*—a term created early in the twentieth century by the German biologist Jakob von Uexküll. You can think of an animal's umwelt as its sensory world. Because their sensory apparatus varies, different species may have different perceptions of the world even if they inhabit the same environment.

For instance, owls, bats, and moths all fly in the nighttime, yet differences in their biology predict differences in the umwelt for each. Owls rely mainly on vision and hearing to catch their prey. Bats also depend on hearing but in a way that is quite different from the owl: they interpret echoes of their own high-pitched calls, hunting and navigating by echolocation. Moths, as invertebrates, may be the least relatable of the three from the perspective of our own umwelt, but we do know that they have good vision and that they can find mates over long distances with their superb perfume detectors. How a species' senses work goes some way toward understanding the mysteries of its felt experiences.

We can expect fishes' umwelts to differ from ours since they evolved in water and not air. But evolution is a conservative designer with a tendency to hang on to a neat idea. Case in point: fish eyes. Apart from their obvious lack of eyelids, fishes' eyes resemble our own. Like most vertebrate eyeballs, including humans', a fish's eyeballs are served by three pairs of muscles that swivel the eye on all axes, as well as a suspensory ligament and retractor muscles that help the fish focus on those bubbles dancing up from the aerator, or that upright creature staring intently from the other side of the glass. As the evolutionary forebears of land-dwelling animals, early fishes originated this system of seeing. It is not easy to spot the swiveling eye movements of most small fishes, but peer closely next time you visit an aquarium and you should be able to detect eye movements in the larger individuals as they shift their gaze to look at different parts of their surroundings.

With a spherical lens of high *refractive index*—defined as the ratio of the speed of light through a medium (in this case, the lens) to its speed through a vacuum—a fish can see as clearly underwater as we can see in air. Needless to say, fishes have neither lacrimal glands nor tear ducts, nor eyelids to moisten the eyes' delicate surface; they don't need them, since the eyeball is kept constantly clean and moist by the water they swim in.

Seahorses, blennies, gobies, and flounders have further upgraded their eye musculature to allow each eye to rotate independently, as in chameleon lizards. I can only conclude from this that a creature thus endowed is able to process two visual fields at a time. That seems just so radically different from what human brains do, and when I try to imagine the mental experience of two independent visual fields, each under my conscious control, it exceeds my umwelt no less than trying to imagine a limit to the universe. Although a team of scientists from Israel and Italy have simulated the visual system of chameleons by building a "robotic head" with two independently moving cameras, I am not aware of any attempts to understand how a single brain processes them. Is a chameleon having two thoughts simultaneously as one eye focuses on a juicy grasshopper on the neighboring twig while the other eye searches the branches overhead for a better approach route? Can a seahorse ogle a potential mate with one eye while tracking the movements of a lurking predator with the other? My single-track brain can't. If I read the newspaper while the radio plays *This American Life*, my mind can toggle back and forth between the two, but try as I might I cannot stream both stories at the same instant.

I also have trouble getting my head around the visual experience of flounders, especially during their early childhood. Baby flounders look like any other normal fish, swimming upright with one eye on each side. Then, in preparation for adult life, they undergo a bizarre transformation: one eye migrates to the other side of the face. It's like facial reconstructive surgery, only in slow motion, and without scalpels and sutures. It isn't even always slow. The entire migration takes just five days if you're a starry flounder, and less than one day in some species. If a fish can have an awkward adolescence, this one qualifies.

In exchange for the indignity of having both eyes nestled next to each other on one flank, flounders have fabulous binocular vision. Like proud neighbors, the two eyes protrude from the body, and each can swivel independently. (Could it be that flounders are the only fishes able to startle themselves by looking themselves in the eye?) Binocular vision is a useful adaptation for a lifestyle of lying in wait on the sandy or stony bottom, exquisitely camouflaged against the substrate, watching for an opportunity to snatch an unsuspecting shrimp or other unfortunate passerby with a lightning-fast lunge. With refined depth perception, a flounder can better judge the timing and wisdom of her ambush.

Ocular migration has obviously proven an effective survival strategy for flounders and related flatfishes, of which there are more than 650 species, including soles, turbots, halibuts, sand dabs, plaices, and tonguefishes. Some species are referred to as "right-eye flounders," always lying on their left side after their left eye migrates to the right side of the body. Others are lefteye flounders. Despite their fine adaptations, many Atlantic flounder and sole species are now threatened by overfishing.

The four-eyed fish—which inhabits fresh and brackish waters along the Atlantic coast of Central and South America—enhances its visual field in a different way. Nature's inventors of the bifocal lens, these relatives of the guppy sport a discrete demarcation between the upper and lower portion of their retina. The fish swims so that the demarcation aligns exactly with the plane of the water surface, the airborne portion of the eyes providing ideal air vision while the submerged portion accommodates the aquatic medium. Flexible genetic coding makes the upper eyes sensitive to green-light wavelengths that predominate in air, and the lower eyes more sensitive to the yellow wavelengths found in muddy waters. This is a valuable visual tool kit when you want to search for a tasty morsel below without being surprise-attacked by a hungry bird from above.

Most larger, faster, open-ocean predatory fishes, including swordfishes, tunas, and some sharks, rely on speed and keen vision to catch prey. The eyes of a twelve-foot swordfish can measure nearly four inches across. Yet hunting underwater presents a set of special visual challenges. If you've ever entered a cave without a flashlight, then you'll have a sense of what fishes experience as they dive deeper below the surface, where there is less light available to see with. There's another problem: water temperatures drop with greater depth, and cold retards brain and muscle function, delaying response times.

To surmount the sluggish effects of cold, some fishes have evolved an ingenious means of improving the functioning of their brains and eyes: they harness heat generated by their muscles, which powers their sensory organs to perform at higher capacity. Swordfishes can heat up their eyes twenty to thirty degrees Fahrenheit above the water temperature. The heat is generated by a countercurrent exchange between the incoming and outgoing blood vessels surrounding the eye muscles. Arteries bringing cold blood from the heart and veins are warmed by a special heat-generating organ in one of the eye muscles. These arteries form a tight, latticed network, enhancing the exchange of heat between them. Studies of eyes removed from recently caught swordfishes suggest that this warming strategy improves by more than tenfold the fish's ability to track rapid changes in prey movements.

Unlike swordfishes, many sharks prefer to hunt at nighttime, when light levels are exceedingly low. Supremely adapted to their realms, sharks' eyes have a layer of reflective cells called the *tapetum lucidum* (Latin: "bright tapestry") next to the retina. Light hitting this layer bounces back through the shark's eye, striking the retina twice and effectively doubling the shark's night vision. This effect is what creates the familiar "eyeshine" of cats and other terrestrial night stalkers. If sharks walked on land, you would see them in the headlights at night by the eerie glow of their eyes.*

Avoiding predators is no less a priority than is catching prey. Be it in an ocean, lake, or stream, fishes use a variety of visual

*There are "walking sharks," but they prefer walking on the ocean bottom to dry land.

techniques to get the upper fin. For those living in the shallows, for example, the underside of the water surface acts as a mirror. This enables a fish to see the reflection of objects that are not in direct view. A bluegill—a saucer-size fish that lives in the shallows of North American lakes, ponds, and slow-moving streams—may be able to spy on a predatory pike lurking on the far side of a rock or thicket of pondweeds by gazing up at the surface reflection. What's good for the goose is also good for the gander, and I'd expect that predators may also use this technique to spy on their prey. I believe this could be studied quite easily in a temporary captive setting.

The mirror technique that the bluegill uses only works in calm waters, and in such conditions fishes can also see quite well what is going on above the surface, allowing them to take evasive action when a diving bird approaches. The fact that wavy water impairs the ability to resolve objects above the surface might explain why seabirds hunt more often and catch more fishes in wavy than in calm waters. The refractive properties of calm water also enhance fishes' ability to see objects on the shoreline. Fishermen armed with this knowledge sometimes stand farther away from the water's edge to reduce the likelihood of detection by their quarry.

Color Badges and Flashlights

There are times, of course, when being detected is the goal. Coral reefs present diverse opportunities for visual innovation. Corals grow in tropical seas at shallow depths, where temperatures and light levels are high. Light does magical things with color, which accounts for the mesmerizing kaleidoscope displayed on the bodies of reef fishes. In fact, when scientists in 2014 discovered evidence of rods and cones in a fossilized sharklike creature that lived 300 million years ago, they concluded that color vision was invented underwater.

In the ages since, fishes have evolved visual capacities beyond our own. For example, most modern bony fishes are tetrachromatic, allowing them to see colors more vividly than we do. We are trichromatic creatures, which means we possess only three types of cone cells in our eyes and our color spectrum is more limited. Having four types of cone cells, fishes' eyes provide four independent channels for conveying color information. Some fishes also see light in the near ultraviolet (UV) spectrum, where light's electromagnetic wavelengths are shorter than what we can see in the so-called "visible spectrum." This helps explain why about one hundred known species from twenty-two families of reef fishes reflect large amounts of UV light from their skin. It all makes me wonder whether a fish gets more excited to see a diver whose wetsuit has blue and yellow racing stripes compared to one wearing a plain black wetsuit.

In 2010, scientists made a discovery that illustrates the value of having a wider visual spectrum than someone else has. Their work focused on visual communication in damselfishes—a colorful and diverse group of reef denizens. They studied two species—the ambon damselfish and the lemon damselfish—which inhabit the same reefs in the western Pacific, and which, to humans, look identical. Ambon damselfishes defend their territories most vigorously against members of their own species. But how do they know an intruder isn't merely a lemon damsel? The researchers had a hunch that vision was still somehow playing a role. It turns out each species has a different facial pattern visible only in the UV light spectrum. When researchers shone a UV light on them, the damsels' faces revealed attractive patterns of dots and arcs resembling a fingerprint, which differed between species in a subtle (to humans) but consistent way. Tested for their recognition skills in captivity, the fishes could reliably indicate correct choices by tapping a picture of their own species with their mouths in return for a food reward. When the researchers used UV filters to eliminate this visual information, the fishes started failing the tests. Furthermore, because the predators of damselfishes appear blind to UV light, the damselfish's face recognition system operates covertly without compromising the camouflage that helps them avoid being seen by their finned foes. It's like being the only one to know who's behind that alluring mask at the masquerade ball.

Fishes' bodies have a variety of ways of expressing themselves through color. In addition to species identification, the coloration of many fishes conveys information to their species-mates about gender, age, reproductive status, and mood. Pigmented cells in the skin contain carotenoids and other compounds that reflect warm colors: yellow, orange, and red. White coloration is not produced passively, by a lack of pigment, but actively, by light reflected from crystals of uric acid in *leucophores* (from the Ancient Greek: leukos=white) and guanine in *iridophores* (iridescent *chromatophores*). Greens, blues, and violets are mostly produced by structural patterns in a fish's skin and scales, and further varied by the thickness of these tissues. Think of a very colorful "clownfish" (such as the Disney character Nemo), whose coloration identifies him as a particular species of anemonefish, and signals a conspicuous warning to other fishes that it would not be a good idea to follow him into the stinging tentacles of his home anemone.

If wearing bright clothes is useful, being able to change them may be even better. By expanding or contracting their melanophores—clusters of cells containing black granules—fishes like cichlids and boxfishes are able to quickly turn darker or lighter in color. Some fishes, such as flounders and cornetfishes, have remarkable control over which cells expand or contract, while colorful coral-reef fishes in particular can usually control the intensity of their so-called "poster coloration." They can ramp up their beauty to lure a potential mate or intimidate the competition, or tone it down to mollify an aggressive competitor or go undetected by a predator.

I think of the flatfishes (the ones with the migrating eyes we visited earlier) as the champions of pigment manipulation. They use their skin to melt chameleonlike into the background. I remember flipping through a biology textbook when I was in high school and encountering a jaw-dropping photo of a flounder who had been placed on a checkerboard in his tank. Within minutes, the flatfish had produced a fine rendition of a checkerboard across his back. From a distance, the flounder effectively disappears. This ability to mimic backgrounds by changing the distribution of skin pigments is a complex and poorly understood process that involves vision and hormones. If one of the flounders' eyes is damaged or covered by sand, they have difficulty matching their colors to their surroundings, which hints at some level of conscious control by the flounder rather than a cellular-level mechanism.

Surrounded by friends and enemies, fishes face a compromise between being detected and not being detected. Near the surface in the Sunlight Zone, practically everything is visible. But light penetration in water decreases exponentially with depth. Being seen is a high priority for a fish, for 90 percent living in the Twilight Zone between 100 and 1,000 meters (0.6 miles) have light-emitting organs (*photophores*) that serve as beacons in the darkness. The proportion is even higher for fishes living in the Midnight Zone, the vast abyss at depths of 2,000 meters and beyond, where no light reaches. Fishes who live here include bristlemouths, lanternfishes, and the famous anglerfishes.

Down here, most of the light is produced by luminous bacteria that coexist with the fishes in an ancient symbiosis. In return for room and board, the light-producing bacteria provide a range of benefits to their hosts. Deep-sea anglerfishes are experts when it comes to light displays. They emit light from the fishing lure that protrudes from their head, and in some species also from a treelike structure suspended from the lower jaw. These glowing adornments enhance their attractiveness to potential prey who, drawn like moths to a candle, swim to their deaths in the jaws of these ambush predators. On the flip side, sudden bursts of light cast from the same structures can be used to startle would-be predators. Body lights can also provide camouflage by casting a faint glow across the fish's lower side, making her less visible against the dim light filtering down from above. And when fishes want to spend some time with companions, the distinctive light patterns produced by these organs can help in recognizing others of their own kind.

Ponyfishes have a peculiar method of luminescence. The *photophore*, or package of light-producing bacteria, carried by males around their throat is shone inward toward a specialized swim bladder (a gas-filled organ that helps control buoyancy) with a reflective coating. The light bounces off this coating and out through a transparent patch of skin. By controlling a muscular shutter in the body wall, the ponyfish creates a flashing display. Schools of males sometimes coordinate their flashing to create a dazzling show, which scientists believe is a strategy to get females into a mating mood.

Flashlight fishes—one of the few bioluminescent fishes generally not found in deep waters—take a more direct approach to illumination, using a multifunctional light consisting of a semicircular organ just below each eye. This pair of organs contains luminescent bacteria whose continuously emitted light can be turned on and off by the fish using a muscular lid. Like ponyfishes, flashlight fishes gather in nighttime shoals, where their combined light helps attract, as well as illuminate, zooplankton prey. These fishes also use the light to evade predators. When the danger approaches, the targeted fish keeps his light on until the last moment before switching off and changing direction. (That must take some nerve.) Mated pairs of flashlight fishes maintain territories over a reef, and if an intruding flashlight fish approaches, the female of the pair will swim up and flash her light literally in the interloper's face, as if to say: "Get lost!"

These deep-sea light shows happen in the blue-green spectrum, which is the color of most bioluminescence, probably because aquamarine light travels farthest through water. But there is one group of fishes that breaks the color rule: the loosejaws. Named for a capacious lower mandible whose flexible hinge allows an enormous gape, these fishes might as well have been called the stoplight fish (actually, one of them is), for the powerful beam of light they shine from a concentrated photophore beneath each eye is red. The color is achieved by a unique fluorescent protein in some species and by a simple gel-like filter over the photophores in others. Naturally, evolution has seen to it that loosejaws can see red, thanks to a small change in a gene responsible for eye pigment structure.

The advantage is huge: a flashlight beam that only the bearer can see. Thus endowed, these hunters of the abyss can spy on others without being seen. Whereas other deep-sea fishes use their lights intermittently, flickering and flashing lest they be discovered and eaten, loosejaws audaciously keep their lamps lit full-time, invisible to their predators and to the prey they stalk with impunity. It is the deep-sea answer to night-vision goggles.

Fooled You!

Clearly, fishes have a diverse and innovative visual repertoire. Their tools are used to enhance their seeing ability, to make themselves more or less visible, to declare their identities, to lure and repel, and to manipulate.

But how do fishes perceive what they themselves see? What is the mental experience of a fish, and how might it compare to our own?

One way of probing this question is by considering optical illusions. If an animal is unaffected by a visual image that fools us, then it would seem that that animal perceives visual fields in a mechanical way, as a robot might "perceive" them. If, however, they fall for the illusion as we do, it suggests that they have a similar mental experience of what they are seeing.

In *Alex & Me*, Irene Pepperberg's touching memoir of thirty years with an African Grey parrot, one of the many captivating findings reported is that these intelligent birds perceive optical illusions as we do: they are fooled by them. The implication, as Pepperberg remarks, is that parrots *literally see the world as we do.*

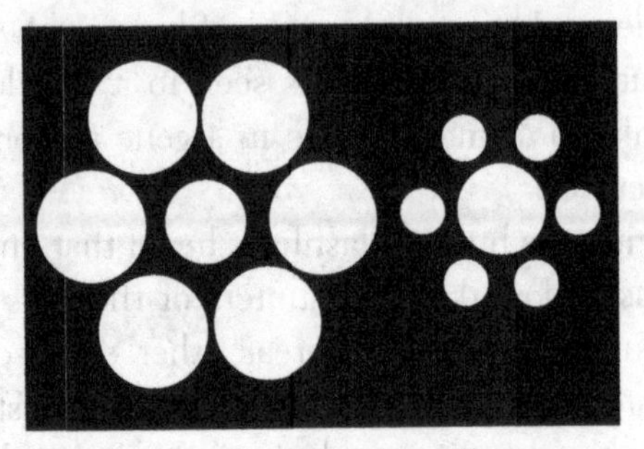

Figure 1: The Ebbinghaus illusion.

Are fishes fooled by optical illusions? Well, in a captive study of redtail splitfins—small fishes that originate from highland Mexican streams—they learned to tap the larger of two disks to get a food reward. Once they had mastered the task, the scientists presented them with the Ebbinghaus illusion, which consists of two disks of the same size, one of which is surrounded by larger disks, making it appear smaller (at least, to human eyes) than the other, which is surrounded by smaller disks (see Figure 1). The splitfins preferred the latter disk.

This result showed the scientists that redtail splitfins do not perceive things in a mindless, stimulus-response way. Rather, they form mental concepts—sometimes fallible ones—based on their perceptions. Similarly, an earlier study found that redtail splitfins also fall for the more familiar Müller-Lyer illusion, in which two identical horizontal lines appear to have different lengths (see Figure 2). Trained to choose a longer line, they chose the line labeled B.

Studies of goldfishes and bamboo sharks show that they, too, respond to visual illusions. You can train goldfishes to discriminate black triangles from black squares on a white background. Then

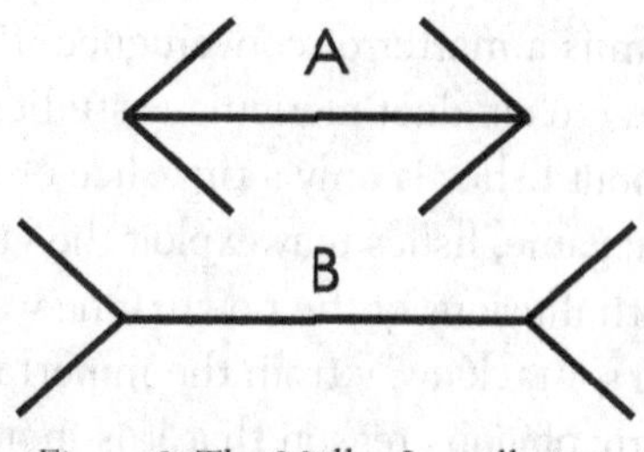

Figure 2: The Müller-Lyer illusion.

if you present them with a Kanizsa triangle or a Kanizsa square, they perceive a triangle and a square, respectively. Kanizsa illusions were developed in the 1950s by the Italian psychologist Gaetano Kanizsa. When humans regard these figures we see a white triangle (or a white square) that looks slightly brighter than the background, even though no triangle is actually drawn (see Figure 3). So what the goldfish brains are doing is the same as what ours do—completing an incomplete picture.

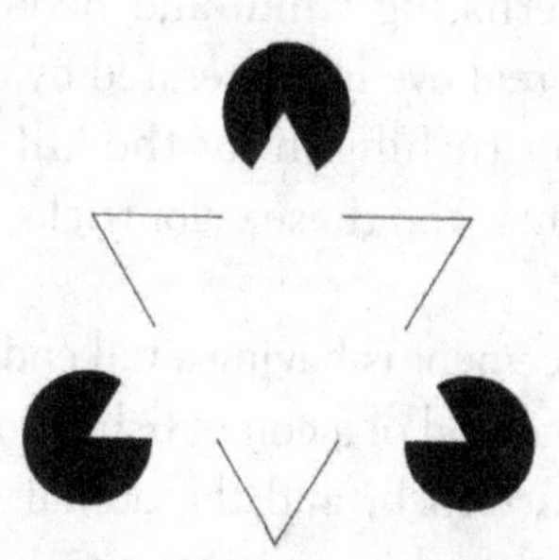

Figure 3: A Kanizsa triangle.

That splitfins, goldfishes, and bamboo sharks can complete an incomplete picture isn't meant to imply that they are unique among fishes in falling for optical illusions. They were simply the species chosen for these studies. Splitfins and goldfishes are only distantly related, so it seems likely that many other fishes would be fooled by optical illusions. These species are studied for the mundane reason that their care in captivity has been well worked out, so using them is a matter of convenience. It takes time and effort (and money) to conduct meticulous studies of animals. So, what we know about fishes is only a tiny slice of what *they* know.

In the survival game, fishes may exploit the visual perceptions of other fishes with illusions of their own. One way to do this is to deflect a predator's attack away from the important parts of one's body. For the fairly obvious reason that it is more likely to be lethal, predators usually direct an attack at the head end of their prey. That many aquatic predators tend to aim for the eyes is evidenced by the evolution of deceptive eyespots on many fishes. Examples of fishes who benefit from this deception include cichlids, butterflyfishes, angelfishes, pufferfishes, and bowfins. The deception may be enhanced in various ways. Like us, fishes are more likely to notice bright colors, so those deceptive eyespots tend toward conspicuous brilliance, while the real eye at the other end may be relatively obscure. The pattern of a young emperor angelfish doesn't include an eyespot, but a bull's-eye surrounded by concentric rings of alternating white and neon blue looks just as effective, while the real eye is obliterated by a maze of meandering lines. A predator rushing in for the kill won't have time to make fine assessments, and these color tricks might tip the scales in the prey's favor.

A further enhancement is having a tail end shaped to resemble a fish's head. The rear end of a comet fish is so arranged to resemble the face of a parrotfish, and the actual eye is virtually lost amid a constellation of white spots covering the entire body, including the eye itself. A behavioral manipulation can enhance these effects even further. Scientists have observed two species of butterflyfish who switch gears and swim slowly backward at the first sign of trouble, then suddenly lurch into forward overdrive if a predator darts in. If they move fast enough, the predator may be snapping at empty water. Otherwise, a butterflyfish is more likely to live on if the chunk of missing flesh came from the tail than if it came from the head.

I find it endearing that fishes perceive optical illusions as we do, and that they are tricked by the visual deceptions of their intended prey. It says something special about the perceptual world—the umwelt—of another being that her mind should construct something that isn't actually there. It suggests the capacity for belief. Beliefs and perceptions can be exploited, and as we've already seen (and will see ahead), fishes use a range of deceptions—visual and otherwise—to improve their chances at success.

As highly visual creatures, we may be able to relate to the importance of having the keen vision that most fishes have. From childhood games we know the disorientation of being blindfolded, and we look on with admiration at how well blind humans learn to cope with the challenges. It is doubtful that a blind fish would live long, even if he inhabits the Midnight Zone, where built-in lights rule supreme. But fishes are not solely dependent on seeing to make their livelihoods. Like us, they have evolved other senses to help them navigate life's demands.

What a Fish Hears, Smells, and Tastes

**The universe is full of magical things,
patiently waiting for our wits to grow sharper.**

—Eden Phillpotts

Just as water influences the dynamics of vision, so it does for hearing, smell, and taste. Water is a superb conductor of sound waves, where they are almost five times longer than in air, as sounds travel five times faster in water. Fishes have benefited from this since the dawn of bones and fins, using sound for both orientation and communication. Water is also an excellent medium for diffusing water-soluble chemical compounds, and is well suited for the perception of smells and tastes. Fishes have separate organs for smelling and tasting, although the distinction is blurred because all substances are encountered in a water solution.

As they did color vision, fishes probably invented hearing. Despite the common assumption that fishes are silent, they actually have more ways of producing sounds than any other group of vertebrate animals. None of these methods involve the main method of all the other vertebrates: the vibration of air against membranes. Fishes can rapidly contract a pair of vocal muscles to vibrate their swim bladder, which also serves as a sound amplifier. They have the options of grating their teeth in their jaws, grinding additional sets of teeth lining their throat, rubbing bones together, stridulating their gill covers, and even—as we'll see—expelling bubbles from their anuses. Some land-dwelling vertebrates get creative in producing nonvocal sounds, such as the drumming of woodpeckers and the chest pounding of gorillas, but fishes' terrestrial cousins possess just two types of vocal apparatus—the syrinx of birds and the larynx of all the rest.

With their versatile acoustic portfolio, fishes produce a veritable symphony of sounds, especially in the percussion section. Among the descriptors we have assigned to them are hums, whistles, thumps, stridulations, creaks, grunts, pops, croaks, pulses, drums, knocks, purrs, brrrs, clicks, moans, chirps, buzzes, growls, and snaps. So notable are the sounds of some fishes that we have named the fishes accordingly: grunts, drums, trumpeters, croakers, sea robins, and grunters. Having ears evolved for processing vibrations in air and not water, we were until recently deaf to most of the sounds fishes were making. It was only in the past century, as underwater sound-detecting technology improved, that the list of acoustic fishes began to grow.

And yet, as recently as the 1930s, scientists believed that fishes were deaf. This prejudice probably arose from the fact that fishes lack an external hearing organ. With our human-centric view of the world, such a lack could only mean one thing: no hearing. Now we know better: fishes don't need ears, thanks to water's incompressibility, which is why water is an excellent conductor of sounds. It is not until we peer inside a fish that we find structures modified and recruited for producing and processing sounds.

Karl von Frisch (1886–1982), the Austrian biologist famous for his discovery of the dance language of honeybees, was also a devoted student of fish behavior and perceptions. Decades before he became the corecipient of the Nobel Prize in 1973 for his contributions to the emergence of ethology (the science of animal behavior), von Frisch was the first to demonstrate hearing in fishes. In the mid-1930s, he devised a simple but ingenious study in his lab with a blind catfish named Xaverl. He did this by lowering a piece of meat on the end of a stick into the water near the clay shelter in which Xaverl spent most of his days. Having an excellent sense of smell, Xaverl would soon emerge from his hiding place to retrieve food. After a few days of this routine, von Frisch began to whistle just before delivering the food. Six days later, he was able to lure Xaverl from his lair just by whistling, thereby proving the fish could hear him. This experiment, and others that followed, were critical to advancing our appreciation for the fish's umwelt.*

Xaverl belongs to an evolutionarily successful group called the Otophysi, which number about 8,000 species (including carps, minnows, tetras, electric eels, and knifefishes). They have evolved a specialized hearing apparatus called the Weberian ossicles, named for its discoverer, the nineteenth-century German physician Ernst Heinrich Weber. The ossicles are a series of small bones derived from the first four of a fish's vertebrae behind the skull. These bones have become separated from their parent bones, forming a chain linking the gas-filled swim bladder with fluid-filled spaces surrounding the inner ear. The apparatus aids hearing by acting as a conductor and amplifier of sound waves, in a manner similar to the middle ear ossicles of mammals.

There are ways in which fish hearing surpasses our own. Most fishes hear in the range of 50 hertz (Hz) to 3,000 Hz, which is within our own range of 20 Hz to 20,000 Hz. But careful studies in captive and wild settings have now documented sensitivity to ultrasounds in the upper range of bat hearing: up to 180,000 Hz

*The first description I read of von Frisch's experiment implied that the catfish was already blind from natural causes, but I later learned that von Frisch blinded Xaverl by surgically removing his eyes for the experiment. Von Frisch may have felt some guilt over this, for he named the fish, and referred in his autobiography to his efforts to "make the aquarium comfortable for the little blind fellow."

in American shad and Gulf menhaden. That is well above the upper human limit. It is believed to be an adaptation for eavesdropping on the ultrasonic sounds produced by dolphins, who prey on these fishes.

At the other end of the hearing spectrum, fishes such as cods, perches, and plaices are responsive to infrasounds as low as 1 Hz. Nobody knows for sure why these fishes have evolved the ability to tune in to super-low sound, but the vast aquatic environments they live in provide clues. Water does not move randomly in oceans and large lakes. Global climate patterns generate currents, local weather patterns produce waves, and our moon's gravitational pull drives the constant heave-ho of ocean tides. Moving water also bumps up against cliffs, beaches, islands, reefs, coastal shelves, and other submerged barriers. All of these forces combine to create ambient infrasound. Biologists from the University of Oslo, Norway, think that fishes use this acoustic information for orientation during migrations. Think of it as a fish's equivalent of birds' use of celestial cues. Pelagic (open ocean) fishes may also detect changes in the surface wave patterns caused by distant land formations and different water depths. Sensitivity to infrasound has also been reported in some cephalopods (octopuses, squids, and others) and crustaceans—further evidence of its utility.

Fishes' hearing sensitivity makes them vulnerable to human-generated underwater noise. For instance, the delicate hair cells lining the inner hearing apparatus become severely damaged by high-intensity, low-frequency sounds produced by air guns used in marine petroleum exploration. Intense noise produced by seismic air-gun prospecting off the coast of Norway decreased the abundance and catch rates of cods and haddocks in the adjacent area.

Some fishes can also detect rapid pulses of sound, discerning as individual beats what we hear only as a constant whistle. And they are proficient at sound directionality, distinguishing sounds from directly ahead versus directly behind, and from directly above versus directly below—perceptual tasks that our brains are less able to manage.

That said, 99 percent of airborne sound energy is reflected off the water surface, so fishes—even if they're congregating close to the shore—are not likely to hear, say, a group of humans talking on a beach. However, airborne sounds transmitted through a solid object, such as an oar bumping against the side of a boat, are easily detected by fishes. This is why anglers sitting in a boat learn to be quiet, and why experienced shore fishermen wander a few yards inland before moving to a new spot; they know that the fish they're after may detect vibrations transmitted through the ground.

With ingenuity, we can hear *them*, too. Fishermen along the Atlantic coast of Ghana use a special paddle as a sort of tuning fork. By placing his ear against a paddle immersed in the water, an experienced practitioner can hear the grunts and whines of nearby fishes, and by rotating the flat plane of the paddle he can locate their whereabouts. A fish's keen hearing may also work in the angler's favor, for many fishes may not realize that the worm they hear up ahead is, unfortunately for them, wriggling on a hook.

Whereas migration and predator avoidance are useful functions for fishes' listening, most sound production has a social function. Here's an example from piranhas. When the biologists Eric Parmentier from the University of Liège, Belgium, and Sandie Millot from the University of Algarve, Portugal, placed hydrophones in a tank holding captive red-bellied piranhas, they recorded a variety of sounds, three of which are common enough to have been ascribed possible functions. One, a repetitive grunt or bark, appears to signal a challenge to others. Another, a low thud, is usually made by the largest fish in the group during aggressive behavior and fighting. These two sounds are produced by a fast-twitching muscle next to the swim bladder that contracts 100 to 200 times per second. A third sound occurs when a piranha grinds or rapidly snaps her teeth while chasing another fish. These descriptors hint at a mean-spirited animal, befitting the piranha's pugnacious reputation as a savage devourer of living victims. In fact, piranhas are mostly scavengers, and pose little danger to humans.

Given that fishes use sounds to communicate with one another, might they also use sound to communicate with us? I know of no scientific study to test this, but there are many anecdotes. Karen Cheng, a computer scientist from the Washington, D.C., area, has four rescued goldfishes in a twenty-gallon tank who, she claims, communicate with her at mealtimes. Around feeding time, when Karen or her husband is in the room but not paying attention to them, their goldfishes rise to the surface and make loud smacking sounds with their mouths. They will also hurl their bodies and bang their tails against the aquarium wall, apparently to get their humans' attention. The sounds produced can be heard from the other side of the room. They stop doing it when someone approaches the tank: "They seem aware of us," says Karen. "Whenever we come up to the tank they abort their activities and swim over to the glass. They don't ignore you like the aquarium fish at the doctor's waiting room."

Sarah Kindrick, a clinical protocol administrator with the National Institutes of Health, saw similar behavior in an eight-inch pinktail triggerfish who lived with her for about three years. Furchbar, as she named him, would take a pebble in his mouth and rap it on the glass wall of his tank around the time she would normally feed him. This is not just an example of interspecies communication by a fish, it is tool use (we'll see more on tool use soon).

Concerto in D Major for Fish

A further testament to fishes' acute hearing is their ability to discriminate tonal patterns of sound—specifically, music. Ava Chase, a research scientist at Harvard University, was interested to see if fishes could learn to categorize sounds as complex as music. She conducted an experiment using three pet store–bought koi named Beauty, Oro, and Pepi. Chase set up a sophisticated apparatus in the fishes' tank that included a speaker at the side for presenting sounds, a response button on the bottom that fishes could push with their bodies, a light that signaled to the fish that his response had been recorded, and a nipple near the surface that dispensed a food pellet when the fish swam up and sucked it after a "correct" response. She then trained the fishes by rewarding them (with a food pellet) when they responded to a certain genre of music and not rewarding them for responding while another genre emanated from the speaker. She found that the koi were not only able to discriminate blues recordings (John Lee Hooker guitar and vocals) from classical recordings (Bach oboe concertos), but that they could generalize these distinctions when presented with new artists and composers for each genre. For example, once familiar with the blues of Muddy Waters, the koi recognized its commonality with blues artist Koko Taylor, as they did for the classical music of Beethoven with that of Schubert. One of the three fishes, Oro, had an especially good ear, able to discriminate melodies in which timbre cues had been removed; that is, all the notes had the same quality except their pitch and timing.* Chase concludes: "It appears that [koi] can discriminate polyphonic music [playing multiple notes simultaneously], discriminate melodic patterns, and even classify music by artistic genre."

Despite their skill as music connoisseurs, neither koi nor goldfishes were known by scientists to communicate using sounds. (Let Karen Cheng's observations serve as preliminary evidence to the contrary.) So it remains a mystery why a mute fish would have such discerning acoustic skills, though as we saw earlier there are benefits to being able to tune in to ambient sounds in one's environment.

Being able to discriminate subtle (and not so subtle) qualities

*Other vertebrates have shown music discrimination abilities, including pigeons, Java sparrows, and to a lesser extent rats (discussed in Chase 2001).

in music is one thing, but it makes me wonder: What psychological effect might it have on a fish? Do fishes appreciate music, or is it just a neutral stimulus?

A research team from the Agricultural University of Athens decided to investigate. They divided 240 common carps among twelve rectangular tanks and randomly assigned them to three different treatments: no music (the control group, for comparisons with the music groups), Mozart's "Romanze: Andante" from *Eine Kleine Nachtmusik*, and the anonymous nineteenth-century "Romanza: Jeux Interdits," a name it got from its use in the 1952 French film *Forbidden Games*. The track duration for these pieces was 6:43 and 2:50, respectively, and the assigned fishes were exposed to four hours of it a day for 106 days. Music exposure was done on weekdays only; like office workers, the fishes got the weekends off (probably because the scientists did).

Fishes in both music groups grew faster than the control group. Feeding efficiency (growth per unit of food), growth rates, and weight gains were higher with either of the two romantic music recordings than without, and intestinal function appeared to be improved. When these fishes were presented with noise or with nonmusical human sounds, the research team found no such changes.

It is a central challenge of animal studies that the subjects cannot report to you in plain language (that we understand) how they are feeling. With these data we can only speculate that carps are responding positively, or negatively, to the music. For instance, a skeptic might suggest that the fishes grew stronger from trying to *escape* the incessant sound of violins and oboes. I must say that, much as I enjoy classical music, hearing the same track over and over is not my idea of acoustic heaven.

We should also consider the possibility that the fishes' growth was no reflection of any subjective experience, but a mechanical response to a physical stimulus. An earlier study by the same Greek scientists noted favorable responses (raised appetites and digestive function) in response to Mozart (the only composer used) by gilthead sea bream, a species with very limited, lo-fi hearing. We also should be wary of anthropomorphism, for there may be little basis for assuming that what we perceive as pleasant music is perceived that way by a fish. Perhaps they prefer *any* sound to none at all. On that score, a better control than silence would be a recording of nonmusical sounds.

There are studies dating back a century that find human patients reporting improved relaxation and reduced pain when exposed to music that they enjoy. A 2015 review of 70 clinical trials involving more than 7,000 patients concluded that music was an effective therapy before, after, and even during surgery, and that it reduces patient anxiety and the need for painkillers. My point here is that music—or more generally, patterned, tonal sounds—may tap deeply into our biology with therapeutic results. The implication is that music appreciation might be widespread in nature.

When I asked one of the authors of the Greek studies, the biologist Nafsika Karakatsouli, she expressed uncertainty that carps enjoy music: "I am not at all convinced that music may have substantial positive effects for fish. There is no music underwater! However, there are plenty of other natural sounds, more relevant to fish living underwater, that may have some meaning to fish and may have produced better results. Nonetheless, some of the fish species we examined, especially carp (a species with excellent hearing abilities), did perform better when music was transmitted." Karakatsouli agrees that a better approach would be to see if carps would choose by themselves an environment with music or not.

There isn't anything musical about the sounds that herrings make, but their innovative method might warrant a fish Grammy Award. One paper describes the first example of what might loosely be termed *flatulent communication*. Both Pacific and Atlantic herrings break wind by releasing gas bubbles from the anal duct region, producing distinctive bursts of pulses, or what the research team playfully named Fast Repetitive Ticks (FRTs). A bout of FRTs can last up to seven seconds. Try that at home! The gas probably originates in the gut or the swim bladder. It isn't clear how these sounds function in herring society, but since per capita rates of sound production are higher in denser schools of herrings, a social function is suspected. So far there is no evidence that herrings ever beg your pardon.

I couldn't think of a better segue from fishes' sense of hearing to their sense of smell than herring FRTs. So let us have a sniff at their smell and taste.

A Good (Sense of) Smell

You may think that a dead fish smells bad, but living fishes have a good sense of smell. They use chemical cues (we'll just call them "smells") for finding food, finding mates, identifying danger, and homing. Smells are especially useful in aquatic environments, where murky conditions make vision unreliable. Some fishes can recognize others of their own kind by scent alone. Sticklebacks, for example, use smell to identify mates of their own species, where proximity to another stickleback species might otherwise present the risk of mating with the wrong kind.

The sophistication of the smelling organs of fishes varies greatly, but the basic design is shared among all the bony fishes (the 30,000 or so fish species that are separate from the sharks and rays group). Unlike those of other vertebrates, fishes' nostrils do not do double duty as organs of smell and openings for breathing; they are used exclusively for smell. Each nostril is populated by layers of specialized cells composing the olfactory epithelium, which is folded upon itself to save space, forming a rosette. Some fishes expand and contract their nostrils, and thousands of tiny cilia pulse in sequence to propel water into and out of the sense organ. Signals from the epithelium are sent to the olfactory bulb at the front of the brain.

Smell is an extremely useful sense for some fishes, as evidenced by their legendary sensitivity. A sockeye salmon can sense shrimp extract at concentrations of one part to a hundred million parts water, which translates in human terms to five teaspoons in an Olympic-size swimming pool. Other salmon can detect the smell of a seal or sea lion diluted to one eighty billionth of water volume, which is about two-thirds of a drop in the same pool. A shark's sense of smell is about 10,000 times better than ours. But the champion sniffer among all fishes (as far as we know) is the American eel, which can detect the equivalent of less than one ten millionth of a drop of their home water in the Olympic pool. Like salmons, eels make long migrations back to specific spawning sites, and they follow a subtle gradient of scent to get there.

One of fishes' most useful adaptations is the production of an "alarm chemical" in the presence of danger, such as a predatory fish or a spearfisherman. Once again, we owe it to Karl von Frisch for discovering yet another phenomenon in the world of fish senses. When he accidentally injured one of his captive minnows, von Frisch noticed that other fishes in the tank began darting back and forth and freezing in place—classic predator-evading behavior. Experiments by von Frisch and others showed that injured minnows (among other fish species) release a pheromone—a secreted or excreted chemical factor that triggers a social response in members of the same species. Detecting this particular pheromone causes agitated reactions by the minnows. Von Frisch coined the term *schreckstoff* (which translates literally to "scary stuff") for these pheromones.

The cells that release schreckstoff are located in the skin, and are fragile enough that they will rupture and release the substance if a fish is placed on moist paper. And it is potent stuff: a thousandth of a milligram of chopped skin is enough to elicit a fright reaction from another fish in a 3.7-gallon aquarium. That's like chopping a marshmallow into 20 million pieces, dropping one piece (if you can still see it) into a sink full of water, then trying to taste the sweetness. Schreckstoff must have evolved long ago, for it is produced by several families of bony fishes.

As a freely available signal, schreckstoff acts like a fire alarm that can be used by other nearby fishes, including different species that may also recognize it. Case in point: fathead minnows. When they smell the poop of northern pikes who have fed on other fathead minnows or on brook sticklebacks—both of which produce schreckstoff in their skin—they immediately flee to hiding places or form tighter shoals. But if the pikes have been fed only on swordtail fishes—which do not produce schreckstoff—the minnows show no signs of fear. Thus, it is not the smell of the pike that the minnows are reacting to. Instead, they detect and react to the schreckstoff from the pike's victims. It is probably due to olfactory skills like the minnows' that pikes refrain from defecating in their own hunting grounds.

The schreckstoff reaction illustrates how fishes can extract subtle clues from waterborne chemicals. But schreckstoff is not the only way to detect a fish foe by fragrance. There is the old-fashioned way of simply recognizing the smell of the predator. Juvenile lemon sharks react to the odor of American crocodiles, who sometimes prey on them. If you are an Atlantic salmon, it depends on what your predator has been eating. In a study conducted at Swansea University in Wales, predator-naive juvenile salmons were presented with water containing traces of dung from one of their natural enemies, the Eurasian otter. The salmons only showed a fear response if the otter had been dining on salmon. In those cases they fled the smell, then remained still, and they breathed faster. Salmons exposed to blank water or to dung from otters on a non-salmon diet were unfazed. The scientists concluded that Atlantic salmons apparently do not innately recognize otters as a threat—they perceive them as a danger only if salmon is on the menu. This generalized mode of predator detection works well because it does not require learning the smell of different predators. Instead, one may just learn to recognize who has been eating one's own kind.

If avoiding predators has a rival in the survival game, one candidate is the quest for sex. Just as aromas have been found to play an important role in human sexual attraction, so, too, do sex pheromones get a fish's juices flowing. For one, they help fishes identify who else is in a mating mood. Fishes have the ability to tease out subtle sex cues and use them for personal gain. Experiments from the 1950s showed that male frillfin gobies will start their courtship displays when a sample of water from a tank holding a sexually receptive female frillfin goby is added to their own tank. Later studies show that females are no less perceptive or active in the mating game. Female sheepshead swordtails from Mexico can discriminate the smell of well-fed males from hungry males of their species—two- to three-inch denizens of tropical rapids—and you can probably guess which they prefer: all else being equal, a well-nourished fish is a more resourceful one, which makes him the better sperm donor. Female swordtails do not discriminate the odor of well-fed females from hungry females, suggesting they are responding to male sex pheromones and not merely to food-based excretions.

So far, we've been examining fish sensory systems as separate units, but they need not work in isolation. Male deep-sea anglerfishes illustrate the interplay of senses. They have the largest nostrils relative to head size of any animal on Earth, according to Ted Pietsch, the world's go-to guy on anglerfishes. His book *Oceanic Anglerfishes* is an astonishingly detailed and lavishly illustrated source for everything currently known about these bizarre fishes.

The male anglerfish's nostrils are not his only well-developed senses; his eyes are also well put together, and Pietsch believes the two senses, smell and vision, function in tandem to help males find females in the dark abyss. A female releases a species-specific pheromone, and a male's fine sense of smell helps guide him toward her species-appropriate perfume. This is important because there are at least 162 known species of anglerfishes cruising about in the world's largest habitat, and you don't want to mate with the wrong one. When he gets close enough to the female, the male angler can confirm that she is his type by the signature of light she emits with the aid of glowing bacteria living in her filamentous lure. One can almost imagine a time in the deep past when the god of deep-sea anglerfishes proclaimed, "Let there be light!" and a lot of the guesswork was taken out of finding a mate.

One last note on the olfactory behavior of fishes: it has been widely assumed by the conservative-leaning scientific establishment that fishes' release of chemicals for communication is passive and not consciously controlled, since they lack external scent glands or scent-marking behavior. That's a shaky presumption. Consider a 2011 study of our friends the sheepshead swordtails. In their fast-flowing habitats, males use at least two tactics to make females aware of their pheromones: (1) they urinate more often when they have an audience of females, and (2) when courting they situate themselves just upstream of females.

For better or worse, that implies that in addition to being able to smell a male's sexual readiness, female sheepshead swordtails can also taste it. What else might a fish be tasting?

Tasteful Fishes

For fishes, the sense of taste is used mainly for food recognition. As for all the other main vertebrate groups—amphibians, reptiles, birds, and mammals—the primary organs of taste are taste buds. Fishes also display a range of teeth types, eight in all, including incisors for snipping, canines for stabbing, molars for grinding, flattened triangular teeth for slicing, and teeth fused into beaks for scraping algae off corals.

Like us, fishes have tongues, and gustatory receptors connected to specialized nerves that relay taste signals to the brain. Not surprisingly, most of a fish's taste buds are located in the mouth and throat. But because fishes are quite literally immersed in the medium they smell and taste, many also have taste buds on other parts of their bodies, most often the lips and snout. Taste buds are also more numerous in fishes than in any other animal. For instance, a fifteen-inch channel catfish had approximately 680,000 taste buds on his entire body, including fins—nearly 100 times the human quota. These and other fishes of murky waters taste their way around their environments. (Try as I might, I cannot imagine what it would feel like if my entire body could function as a tongue, but I'm quite sure I would want it to include an "off" switch.) Cavefishes also benefit from a bonanza of taste buds, which provide a high-definition flavor sensing system to aid food finding in the darkness. Many bottom-feeders, including catfishes, sturgeons, and carps, are equipped with barbels, whisker-like feelers usually located around the mouth, which serve as taste sensors.

In case you're wondering why fishes would need a sense of taste—well, they need it for the same reasons we do. Fishes have food preferences that may be distinctive to species and even individuals. It can take a fish a while to ascertain the palatability of a food item; if you've watched fishes in aquariums you might have seen how they will sometimes take a morsel into their mouth, spit it out, then re-ingest it several times before either swallowing or rejecting it. Overall taste preferences within a fish species, and within different populations of the same species, generally do not vary much, as is the case for human ethnic groups. The same does not hold true of individual preferences. In our case, think Brussels sprouts, spicy versus mild, and the dizzying array of present-day variations on a cup of coffee. Studies of rainbow trouts and carps find that finicky eaters are not rare.

Fishes' reactions to unpleasant tastes are reminiscent of our own. Just as we will quickly eject a mouthful (as gracefully as possible if in public) when we bite into a surprisingly rotten piece of fruit or meat, a Dover sole expresses her distaste for a food item by violently turning and rapidly swimming away from it, shaking or nodding her head. Stéphan Reebs, the author of *Fish Behavior in the Aquarium and in the Wild*, describes a fish's reaction to the taste of toad tadpoles—a toxic and particularly foul-tasting item in its environment: "It must be said that a very hungry bass, its back to the wall, will stoop to eat toad tadpoles. But if the reaction of other fishes that mistakenly take tadpoles in their mouth is anything to go by—they violently shake their head and you can almost see the grimace on their face—having tadpoles on the menu is no great culinary experience for a fish."

Living in a relatively dense aqueous medium imposes some limitations, but it also provides fishes with sensory opportunities unavailable to land-dwelling beasts. Can you imagine chatting with your neighbor using electric pulses? In the next part of the book, we'll get beyond the mainstream senses into some less familiar ways fishes use to perceive their environments.

Navigation, Touch, and Beyond

When one flesh is waiting, there is electricity in the merest contact.
—Wallace Stegner

Fishes need to move about to meet their needs, and they must be in certain places at certain times if they are to be successful at making a living and making more fishes. Like us, fishes return to specific places at different times of day, such as feeding locations, hiding and sleeping places, and cleaning stations. At certain times of the year they return to mating, spawning, and nesting sites. Living in a complex volumetric habitat, fishes face a challenging spatial milieu.

Fishes are excellent navigators, and they use a variety of methods to find their way around over both short and long distances. Blind cavefishes live in relatively small cave habitats, but most also live in total darkness, so having good navigation skills is important to them. These little fishes can learn the order of a sequence of landmarks en route to a destination by feeling the turbulence reflected off underwater obstacles. Swordfishes, parrotfishes, and sockeye salmons use sun compassing, setting their direction based on the angle of the sun. Yet others can use dead reckoning—taking wandering, novel outbound exploratory trips from a point of reference and then returning to home base using a direct path.

The navigation feats of salmons are the stuff of legend. Being able to return to their natal streams to spawn after spending years in the open ocean ranks these anadromous fishes (migrating to sea, homing to spawn) as having one of nature's finest built-in Global Positioning Systems. As far as we know, this system uses at least two and possibly three sensory tools to work at full capacity: geomagnetic sense, smell, and possibly vision.

Like sharks, eels, and tunas, these long-distance fishes plug into the Earth's magnetic field to aid their navigation. This manifests at the cellular level. Single cells containing microscopic magnetite crystals act like compass needles. By isolating cells from the nasal passages of trouts (very close relatives of salmons) and exposing these cells to a rotating magnetic field, a research team from Germany, France, and Malaysia found that the cells themselves rotated. The magnetite particles are firmly attached to the cell membrane, and by constantly pulling toward magnetic field lines, these particles generate torque on the cell membrane when the salmon changes direction. That torque must be directly transmitted to stress-sensitive transducers of some kind, because evidence shows that the salmons can feel it.

They also use their prodigious sense of smell. When heading downstream to the ocean, young salmons "record" the water chemistry along the way. Years later, they retrace their paths, following the distinctive odor signature of their home stream, like walking a trail in reverse. Anosmic salmons, whose noses had been experimentally plugged by biologists to eliminate their ability to smell, showed up in random streams, whereas unmolested fishes returned to their home streams to spawn.

In a less invasive experiment, the same research team, led by the late Arthur Hasler of the University of Wisconsin, split a group of young coho (freshwater) salmons into two groups, each of which was exposed to one of two different, innocuous but fragrant chemicals—morpholine and phenylethyl alcohol (PEA).

After this exposure period, salmons from both groups were released together directly into Lake Michigan. During the salmon spawning migration one and a half years later, the researchers dripped morpholine into one stream, and PEA into another located five miles away. Nearly all of the recaptured salmons in the morpholine-scented stream were from the morpholine group, and nearly all of their PEA counterparts navigated up the other stream.

Might a salmon also use vision to aid navigation? A Japanese research team sought to find out in a study involving ocean release and recapture of sockeye salmons. The scientists blinded a sample of the fishes before release by injecting their eyes with carbon toner and corn oil. Upon recapture five days later, only 25 percent of these salmons, compared to 40 percent of the unaltered fishes, were caught in their natal stream. The authors suggested that these fishes are nevertheless using vision to reach the entrance to their natal stream, but I find this result unconvincing. I suspect that the pain, distress, and ensuing disorientation caused by blinding salmon with an injection of foreign substances might explain their lower success rate in finding their way home. To better control for that, they would need to have injected some salmons with a similar amount of solution that doesn't cause blindness. But I'm not recommending it.

Pressure Sensors

Not only do fishes navigate independently, they have another system of orientation that allows them to track closely the movements of their neighbors. Like flocking birds, who use vision and hair-trigger reflexes to coordinate their flight directions with their neighbors, large schools of fishes can change direction seemingly as one, as if there were some internal knowledge of the decision making of all others. It isn't clear who starts it, or if the chain reaction starts with whoever happens to make the first move.

Early naturalists ascribed this behavior to a form of telepathy, but analysis of slow-motion filmed sequences yields a no-frills explanation: minuscule delays in the propagation of movement through the school show that the fishes are reacting to each other's movements. Their sensory systems are operating on such a fine timescale that it gives the impression that they all change direction as one.

In daytime conditions, sharp eyesight helps schooling fishes move in unison as birds do. But unlike birds (or humans who dare try it), they continue to move as one even in darkness. How? It's thanks to a row of specialized scales running horizontally along their flanks, forming what is called the lateral line. The lateral line is usually visible as a thin, dark line because each scale has a depression that casts a shadow. The depression is populated by neuromasts, clusters of sensory cells each with a hairlike projection encased in a tiny cup of gel. Changes in water pressure and turbulence, including waves from the fish's own movement reflected back from its surroundings, cause deflections of the neuromast hairs, which trigger nerve impulses to the fish's brain. So the lateral line acts as a sonarlike system and is especially useful at night and in murky waters.

With the lateral line, fishes swimming in close proximity are virtually in physical contact, and the transmission of signals between them is comparable to that of visual information, giving rise to hydrodynamic imaging. It is hydrodynamic imaging that allows blind cavefishes to detect stationary objects such as rocks and coral by the distortion of the normally symmetrical field flow that surrounds a fish in open water. Blind cavefishes can form mental maps, a skill very useful for navigation by a creature lacking the means for visual orientation.

Lateralization of brain functions is now known to be widespread in fishes, and these clever little fishes also use their lateral lines nonsymmetrically when confronted with unfamiliar objects. When a plastic landmark was placed in their tank along the middle of one wall, blind cavefishes preferentially swam past it using the lateral line on their right side. This preference disappeared in a few hours as the fishes became familiar, and therefore comfortable, with the new landmark. Because visual and lateral line sensory systems operate independently in fishes, this finding suggests that lateralization of the brain is a deep-seated phenomenon. Sighted fishes were already known for a tendency to have a right-eye bias in emotional contexts, such as examining a new (and thus scary) object.

Like most biological designs, the lateral line comes with inevitable compromises. Water flow generated by swimming activates the neuromasts, and this "background noise" dampens the fish's reactivity to external movements. Experiments show that swimming fishes are only half as likely as stationary fishes to respond to the movements of a nearby predator. On the other hand, a fish can detect distortions in the bow wave formed in front of his own nose when swimming forward, and thereby avoid bumping into objects made invisible by darkness or transparency, such as an aquarium wall. It is unfortunate for fishes that this system seems unfit for detecting the presence of a fishing net.

Electrified

Having a sense that allows you to avoid bumping into a wall in the dark is useful, but imagine being able to detect the presence of something on the other side of the wall when you cannot see or hear anything. Enter the world of electroreception.

Electroreception is the biological ability to perceive natural electrical stimuli. It is nearly unique to fishes, the only known exceptions being monotremes (platypuses and echidnas), cockroaches, and bees. Electrical sensitivity is widespread in sharks, skates, and rays. Among the teleosts (the 30,000-plus species of bony fishes), more than three hundred species get a charge out of life, and it must have high value as a survival tool, for it has evolved independently at least eight times in fishes. Its predominance in aquatic habitats relates to water's strong electrical conductance properties as compared to air.

As the term implies, electroreception is a passive use of electrical information. The elasmobranchs are electroreceptive only; they can detect electric stimuli but they do not produce electricity themselves. They perceive it with a network of jelly-filled pores scattered strategically over the head. These pores are called the *ampullae of Lorenzini*, after Stefano Lorenzini, the Italian physician who first described them in 1678. Noting the concentration of black specks that surrounds sharks' snouts like a five-o'clock shadow, Lorenzini peeled away the skin to reveal tubular channels—some as wide as strands of spaghetti—leading to the brain, where they congregate in several large masses of clear jelly.

The function of the ampullae of Lorenzini in electroreception remained a mystery until 1960. They detect subtle electrical changes generated by nerve impulses of other organisms, which propagate efficiently through water. Such is the sensitivity of this system that just the heartbeat of a fish hiding six inches under the sand may be enough to betray its presence to a hungry shark or catfish.

Some bony fishes actively produce their own electrical charges. You have no doubt heard of electric eels. These South American river dwellers can grow to seven feet and forty-five pounds. They are named for their elongated shape and are not true eels, belonging instead to the knifefish family, close relatives of the catfishes. They use low-voltage discharges to help them navigate in their murky habitats by detecting the electric fields that bounce off solid objects. But they are better known for producing stunning electrical discharges up to 600 volts or more. The electric organs are housed in stacked cells within the tail musculature. As in the stacked cells of a battery, electricity can be stockpiled until needed, then, if the eel so chooses, released all at once. This built-in Taser gun can be used to stun or kill prey, or to repel unwelcome intruders.*

* You might be wondering how these so-called strongly electric fishes avoid delivering electric shocks to themselves. They have layers of fatty tissue that help insulate them from the brunt of their own weapons. Nevertheless, they sometimes twitch in reaction to their own shocks.

The voltage power of electrical discharges of electric eels and some other fishes, such as torpedo rays, have earned them the term *strongly* electric fishes. But for me the most interesting use of electricity is reserved for certain *weakly* electric fishes, who use it for the less violent purpose of communicating with others of their kind. Most of these fishes belong to two groups: the diverse elephantfishes of Africa—so-named for their elongated, downward-pointing noses—and the knifefishes of South America, named for their pale coloration and knifelike shape. Like so many fishes with stealth technologies, they inhabit muddy waters, which likely provided the adaptive basis for a novel nonvisual means of communication. They communicate with high-speed electric organ discharges (EODs) of up to 1,000 pulses per second, or 1 kilohertz (kHz), more than twice as fast as an electric eel's pulse rate.

They are adept at interpreting these signals, as illustrated by a species of elephantfish that lives in river and coastal basins of western and central Africa. When the biologists Stephan Paintner and Bernd Kramer from the Institute for Zoology at Regensburg University, Germany, presented them with simulated EODs, the fishes showed an "astounding" ability to discern pulse time differences down to a millionth of a second. This rivals echolocation by bats as the fastest form of communication in the animal kingdom.

By varying the rate, duration, amplitude, and frequency of their EODs, elephantfishes can exchange information about species, sex, size, age, location, distance, and sexual inclination. EODs also communicate social status and emotion, including aggression, submission, and mate attraction, for which signals are crafted into courtship "songs" to serenade potential mates using exotic patterns of chirps, rasps, or creaks. (When you communicate your desire with electricity, being "turned on" takes on added meaning.) They can identify other individuals by their EOD signatures, which are distinct and remain stable over time. Dominant individuals may chase trespassers off their territories when they detect the trespasser's EOD, which probably explains why fishes often deferentially turn off their EOD when swimming through a neighbor's territory. Pairs or groups of fishes also coordinate their EODs, producing "echoes" and "duets." Males will alternate EOD pulses with other males, whereas females will synchronize theirs with investigating males.

It could get confusing when a cluster of elephantfishes or knifefishes are chirping away in close proximity. They deal with it using a so-called jamming avoidance response: if two fishes' discharge frequencies are too similar and might interfere with discrimination, they adjust to enlarge the distinction. Fishes in a social group maintain a 10 to 15 Hz difference from neighbors, ensuring that each individual has a personalized discharge frequency.

Recordings of EOD-producing elephantfishes in the upper Zambezi River suggest that they also use their signals to cooperate. EODs produced by fishes threatened by a lurking predator induce neighbors to join in on what may be an early-warning alert. It benefits all fishes in the neighborhood if predators have low hunting success there. Signals exchanged by familiar neighbors can provide assurances that all is well, thereby avoiding the need for costly defense of territory. Such "dear enemies" also team up as shoaling partners when food gets scarce.

If all this sounds too sophisticated for a fish, it may be time to reassess your perceptions of fish intelligence. Consider also that the elephantfishes have the largest brain cerebellum of any fish, and that their brain-to-body-weight ratio—a highly touted marker of intelligence—is about the same as ours. Much of that gray matter is devoted to electroreception and communication.

There is a cost to using electricity to communicate. Electroreceptive predators could be tuning in. Such is the case with sharptooth catfishes, who hunt in packs during spectacular yearly migration runs up the Okavango River in southern Africa. Most of their diet during this time is a species of elephantfish called the bulldog. They locate the hapless bulldogs by eavesdropping on the bulldogs' EODs. But there's a further twist. Captive studies have found that the EODs of female bulldogs are too short for the catfishes to detect, whereas the males' EODs are ten times longer, and the catfishes can easily notice them. The size distribution of bulldogs found in catfish stomachs indicates that it is the males who are mostly being eaten. In the evolutionary arms race of avoiding being someone else's meal, we may expect male bulldogs to be shortening their EODs.

The Pleasure of Touch

While lateral lines and electric organ discharges are alien to our sensory systems, the sense of touch certainly is not. In exploring this familiar sensation in fishes, I want to connect it to another kind of sensation that we often derive from touch, and one that we rarely consider as being part of the lives of fishes. I am referring to the sense of pleasure.

In his iconic poem "Fish," D. H. Lawrence wrote:

> They drive in shoals.
> But soundless, and out of contact.
> They exchange no word, no spasm, not even anger.
> Not one touch.
> Many suspended together, forever apart.
> Each one alone with the waters, upon one wave with the
> rest.

I love these lines, and I can see what Lawrence means: to my airborne senses there is something lonely about fishes being forever suspended in their heavier, viscous medium.

But writing in the early 1920s, Lawrence didn't have the benefit of knowing what we know today about the lives of fishes. Fishes are not alone. They know each other as individuals and they have preferences for who they hang out with. They communicate through diverse sensory channels. They have sex lives. Contrary to the notion of their being separate, it turns out that fishes are highly sensitive to touch, and tactile communication enriches the lives of many.

While researching this book I was sent a video clip by a puzzled viewer who couldn't understand why a fish—in this case a bright-orange Midas cichlid who looks for all the world like a friendly character from *Finding Nemo*—would return repeatedly to be stroked, picked up, and playfully tossed back into the water by a man.

What would motivate a fish to do this?

The answer, I believe, is that it feels good. Fishes often touch one another in pleasurable contexts. Many court with rubbing or gentle nips. Cleanerfishes curry favor with their valued clients by caressing them with their fins as a means to strengthen the cleaner-client relationship. Moray eels and groupers approach familiar divers and receive strokes and chin rubs.

In an informal survey of public perceptions of fishes, I received unsolicited accounts from eight of a thousand random respondents who described behavior like that of the Midas cichlid we just met. These fishes would allow their humans to pet, touch, hold, and stroke them. The author Cathy Unruh later wrote to me about a Bahamian grouper she calls Larry. Whenever Cathy and other divers descend to his reef, Larry swims over to be petted. According to Cathy, Larry seems to enjoy making eye contact, and checking out the divers' bubbles. He even rolls side to side to be petted properly, as a dog or a pig will do. Today one can find videos of fishes cavorting and in some cases appearing to snuggle with divers, who stroke their bodies gently as if they were the family cat. There are also growing numbers of videos of aquarium fishes swimming repeatedly into the hand of a trusted owner to be lovingly stroked.

The other major group of fishes—sharks, rays, and skates—also show pleasurable responses to touch. The diver Sean Payne described an encounter he had with a juvenile manta ray off the Florida coast. The ray swam up to Payne and rubbed repeatedly against him, leading him in a circular tango that forced her body into his hands:

"As I ran my hands over her skin, her wing tips vibrated like a dog's leg during a particularly good belly scratch," said Payne.

Andrea Marshall, the founder of the Marine Megafauna Association, describes manta rays as strongly curious and interactive with humans. These massive elasmobranchs, who have the largest brains of all fishes, love getting bubble massages from Marshall. She swims beneath them and blows bubbles from her SCUBA regulator. If she stops, the rays swim away, but soon return for more. It's a similar story at the Shedd Aquarium in Chicago, where two of the five zebra sharks in a 400,000-gallon tank like to swim among the staff divers. "I think they like the feel of the bubbles coming out of our regulators," says Lise Watson, Wild Reef collection manager. "During our maintenance dives, if we put our regulators underneath them, they dance around while the bubbles tickle their bellies."

Besides touch, there are many other ways that fishes may derive pleasure. Food, play, and sex spring to mind. And then there's comfort for its own sake. Southern bluefin tunas in the waters of Australia spend hours rolling on their sides, catching the sun's rays. It's not known for sure why they do this. One possibility is that they are sunbathing to raise their body temperature, which in turn helps them swim and react faster, making them more efficient hunters. I expect the warmth of the sun also feels good to a tuna, for pleasure evolved to reward useful behaviors.

Ocean sunfishes are named for their fondness for sunbathing while lying on their sides just beneath the surface. These huge fishes are also parasite hotels, harboring as many as forty different species of external parasites, including large copepods that can reach six inches. The sunfishes queue beneath floating kelp beds, waiting their turn to be serviced by cleanerfishes there. The sunfish at the front floats onto its side to signal readiness.

But some of the parasites are too large to be removed by fishes, and this is when the sunfish turns to a specialist. Floating up to the surface, the giant fish invites gulls to surgically remove penetrating skin parasites with their powerful beaks. Sunfishes have been seen courting the birds, following them around and swimming sideways next to them.

Dare we think the sunfish knows the feeling of relief from a skin irritation, and understands the cause-and-effect of bird and parasite? It is the best-fit explanation I can think of for a wise old creature who may live a century and wander thousands of square miles of open ocean.

To know pleasure is to know pain. Or so it would seem. Yet, despite steady advances in our understanding of the full-bodied lives of fishes, the question of their capacity to feel pain remains a subject of debate. Should it? Let's find out.

PART III

WHAT A FISH FEELS

Your life a sluice of sensation along your sides.

—from "Fish," by D. H. Lawrence

Pain, Consciousness, and Awareness

Water wetly on fire in the grates of your gills.

—from "Fish," by D. H. Lawrence

Do fishes feel pain? While it may seem obvious to some of us that they do based on their appearance, their behavior, and their membership in the group of vertebrate animals, many people believe otherwise. I am only aware of limited opinion research on this question, such as a survey of North American anglers and other recreational fisheries stakeholders, which found that slightly more believed fishes feel pain than believed they do not, and a survey of New Zealanders that had a similar result.

The question of whether fishes experience pain is of cardinal importance—recall those astronomical numbers of fishes killed by humans, from the prologue. Organisms that can feel pain can suffer, and therefore have an interest in avoiding pain and suffering. Being able to feel pain is not a trifling thing. It requires conscious experience. An organism may move away from a negative stimulus without any experience of pain. It could be a reflexive response in which nerves and muscles cause the body to move without any mental engagement. For example, a heavily sedated human patient in a hospital setting with no capacity to experience pain may nevertheless recoil in response to a potentially harmful stimulus, such as exposure to heat or intense pressure. This is due to the actions of peripheral nerves working independently of the brain. Scientists use the term *nociception* to describe a reflex that in itself involves no awareness or pain. Nociception is the first stage in pain sensing—necessary but not sufficient for the experience of pain. It is only when information from nociceptors is relayed to higher brain centers that it hurts.

There are some good reasons to expect that fishes are sentient. As vertebrates, they have the same basic body plan as mammals, including a backbone, a suite of senses, and a peripheral nervous system governed by a brain. Being able to detect and learn to avoid harmful events is also useful to a fish. Pain alerts animals to potential damage that may lead to impairment or loss of life. Injury or death reduces or eliminates an individual's reproductive potential, which is why natural selection favors the avoidance of these dire outcomes. Pain teaches and motivates animals to avoid a noxious past event.

I have an assignment for you that might provide some insight on the question of whether fishes are consciously aware and thus capable of pain. Go to a public aquarium. Choose a tank. Spend five minutes watching the fishes in there. Look long and hard. Look closely at their eyes. Watch the movements of their fins and their bodies, keeping in mind what you now know about their vision, hearing, smell, and touch. Choose an individual. Does he pay attention to other fishes? Do you see any organization to his movements, or does he appear to be just randomly swimming about as if on autopilot?

If you do this, you will usually see nonrandom patterns of behavior. You'll notice a tendency for fishes to consort with others of their own kind. You will see—especially in larger fishes with more easily watched body parts—that their eyes are not locked into a fixed stare, but swivel in their sockets. If you are especially patient and observant, you'll note idiosyncrasies expressed by individuals. For example, one fish might appear dominant over another, giving chase when the subordinate transgresses some social or physical boundary. Some individuals may be more adventurous, others more shy.

When I was little, I didn't pay much attention when gazing at "fish" in a tank. I wasn't looking at other beings—only at swimming creatures with shapes and colors. Gradually I began to watch fishes more closely, and they became more interesting. Now, when I linger in front of the glass wall that separates two universes of life, I notice that there is pattern and structure to their swimming, and organization to their social lives. Even in a small tank, which is a poor substitute for the complexity of a natural habitat, fishes usually have favored areas to swim or rest in.

Fishes are certainly awake, but are they aware? Being aware involves having experiences, taking notice, remembering things. An aware creature is not merely alive; she has a life. This book contains a lot of science that supports fishes being aware. But sometimes a story conveys it better than any amount of science can. Ana Negrón, a physician friend of mine from Pennsylvania, shared this account with me:

> It was 1989. I was snorkeling leisurely back to the sailboat anchored in the crystal-clear waters off the northeast coast of Puerto Rico when a four-foot-long grouper and I caught sight of each other. He was so close I could almost have reached out and touched him. His entire left side shimmered in the sunlight. I stopped flapping my fins and froze. We both remained immobile, suspended barely a foot under the surface, looking into each other. As I drifted with the current, his large eye moved in its socket, locked to my gaze for perhaps half a minute, which seemed an eternity. I don't remember who moved away first, but as I climbed back on the boat let it be known that a fish and a

woman had been aware of each other. Although I have looked into the eyes of whales since, I still feel this fish's presence the strongest.

When I watch what fishes do—swimming through the water, chasing one another, coming to one end of an aquarium to be fed—my common sense emphatically tells me that they are conscious, feeling creatures. It goes against my deepest intuition to think otherwise. But common sense and intuition do not a science make. Let's see what the science says about sentience in fishes.

The Fish Sentience Debate

Two key players in the fish-feel-pain camp are the fish biologists Victoria Braithwaite at Pennsylvania State University and Lynne Sneddon at the University of Liverpool. James Rose, a professor emeritus at the University of Wyoming, denies that fishes feel pain. In 2012 Rose and six colleagues—each with impressive academic credentials—published a paper titled "Can Fish Really Feel Pain?" in the journal *Fish and Fisheries*. The crux of their argument is their belief that fishes are unconscious (meaning unaware of anything, unable to feel, think, even see), and because pain is a purely conscious experience, fishes therefore cannot experience pain. The basis of their claim is what I call *corticocentrism*—the claim that to "possess a humanlike capacity for pain" one must have a neocortex, the cauliflower-like portion of the brain that features ridges and in-foldings. *Neocortex* translates from its Latin roots to mean "new bark," denoting the new layer of gray matter that is thought to be the most recently evolved part of the vertebrate brain. Only the brains of mammals have it.

If the neocortex is the seat of consciousness, and only mammals have one, it follows that all nonmammals lack consciousness. But there is a major snag here. Birds lack a neocortex, yet the evidence for consciousness in birds is virtually universally accepted. The cognitive feats of birds include tool manufacture; remembering for months the locations of thousands of buried objects; categorizing objects according to combined characteristics (such as color and shape); recognizing a neighbor's voice over successive years; using names to call one's chicks back to the nest at sunset; inventive play, such as sliding down snowbanks or car windows; and clever mischief, such as stealing sandwiches and ice-cream cones from unsuspecting tourists. So impressive are the conscious acts of birds that the nomenclature of the proverbial "birdbrain" was overhauled in 2005 to reflect the parallel evolutionary pathway that the avian paleocortex (old bark) has taken—allowing birds to function cognitively at comparable levels with mammals. Birds lay waste to the idea that a creature needs a neocortex to be aware, have experiences, and do clever things. Or feel pain.

If any animal without a neocortex is nevertheless conscious, it disproves the notion that a neocortex is required for consciousness. As such, it is no basis for a claim that fishes are unconscious. "There are many ways to get to a complex awareness," says the neuroscientist Lori Marino of Emory University. "To suggest that fishes cannot feel pain because they don't have sufficient neuroanatomy is like arguing that balloons cannot fly because they don't have wings."

Or that humans cannot swim because they don't have fins.

The fishes' answer to the mammalian cortex is the *pallium*, which is noted for its astonishing diversity and complexity. And while there is less computational power in the average fish pallium than in the average primate neocortex, it is increasingly apparent that the pallium serves functions for fishes that the neocortex does for mammals and the paleocortex for birds. We'll explore these capacities ahead, but for now, let me just mention learning, memory, individual recognition, play, tool use, cooperation, and account keeping.

Returning to the Hook

Let's address the situation in which a fish gets repeatedly hooked in quick succession. "Stories abound of bass that are caught and released, only to turn around and be taken again the same or next day, sometimes more than once," writes fish biologist Keith A. Jones in a book for bass anglers. Some fishermen claim, understandably, that this suggests the experience of being hooked is not traumatic for the fish. Otherwise, why would they so quickly strike bait again? (We might as soon ask why a fish would return repeatedly to the hand of a fisherman to be petted if it couldn't feel anything.)

But "hook shyness" is also a term familiar to most fishermen. There are studies in which long periods of time elapsed before fishes resumed normal activity following capture by hook and line. Carps and pikes avoided bait for up to three years after being hooked just once. A series of tests on largemouth basses showed that they, too, quickly learned to avoid hooks and remained hook-shy for six months.

There are also studies in which fishes resumed what appeared to be normal behavior within minutes of being subjected to invasive procedures such as surgery to implant transponders to track their movements in the wild. I fail to see how this should cast doubt on fish pain. A very hungry fish who is in pain does not cease to be hungry, so the motivation to feed may override the inhibiting effects of traumatic pain.

In a 2014 interview, Culum Brown, who researches fish cognition and behavior in the Department of Biological Sciences at Macquarie University, in Sydney, responded to the repeat-hooking phenomenon:

> They need to eat. There is too much uncertainty in the world to let a meal go by. Many will strike even when they are completely full. . . . People will often say to me, "but I keep catching the same fish." Well yeah, if you were starv-

ing and someone kept putting a hook in your hamburger (say 1 in every 10 had a hook) what would you do? You keep eating hamburgers because if you don't you starve to death.

Pain Studies in Trouts

The matter of hook shyness proves little, and scientists and philosophers will likely continue to debate animal consciousness for a long time to come. For probing fish sentience, we would do better to look at scientific studies on fish pain. A substantial body of research exists on the subject, of which I can provide only a small sample in a book of this scope. Among the most meticulous experiments are those performed on rainbow trouts—a representative bony fish—by Braithwaite and Sneddon. Their findings are summarized in Braithwaite's book *Do Fish Feel Pain?*

The first step in examining the capacity for pain in fishes is to see if they are equipped for it. What sorts of nervous tissue do fishes have, and does it function as we would expect in a sensate animal?

To find out, trouts were deeply and terminally anesthetized (they were knocked out for the duration of the experiment and then killed with an overdose of the anesthetic at the end) and their facial nerves surgically exposed. The trigeminal nerve—the largest of the cranial nerves, which is found in all vertebrates and is responsible for sensation in the face and motor functions such as biting and chewing—was examined and found to contain both A-delta and C fibers. In humans and other mammals these fibers are associated with two types of pain sensation: A-delta fibers signal the sharp initial pain of an injury, whereas C fibers signal the duller, throbbing pain that follows. Interestingly, the researchers found that C fibers were present in a much lower proportion in trouts (about 4 percent) than has been found in other vertebrates studied (50 to 60 percent). This suggests that, in trouts at least, persistent pain following initial injury could be less severe. But the proportion skew may mean little, for, as Lynne Sneddon has pointed out, trout A-delta fibers act in the same way as mammalian C fibers, reacting to a variety of noxious stimuli.

Next, the research team wanted to find out whether noxious stimuli delivered to the trout's skin would activate the trigeminal nerve. This was done by stimulating the trigeminal ganglion, a region where the three sensory branches of the trigeminal nerve converge. Microelectrodes were guided into individual nerve cell bodies in the ganglion, then three kinds of stimuli were applied to receptor areas on the head and face: touch, heat, and chemical (weak acetic acid). All three generated rapid bursts of activity in the trigeminal nerve as registered by electrical signals in the electrodes. Some nerve receptors responded to all three stimulus types, others to one or two. This provided an important clue to the scientists that trouts are equipped to respond to different types of potentially painful events: mechanical injury (like cutting or stabbing), burning, and chemical damage (from acid).

Being equipped to experience pain is a solid foundation for the conclusion that an organism is sentient, but it is not the final word. Even in the face of the evidence accumulated so far, it could still be that the neurons, ganglions, and brains of fishes can only register a negative stimulus in a reflexive way, without any actual sensation of pain.

In the next phase of the experiments, trouts were subjected to one of four treatments: after being netted, then briefly anesthetized, they were either (1) injected in the mouth (just under the skin) with bee venom, (2) injected with vinegar, (3) injected with a neutral saline solution, or (4) similarly handled but not injected. Manipulations 3 and 4 allowed the researchers to cancel out the effects of being handled and injected with a needle. The trouts were then returned to their home tank to be watched from behind a black curtain to avoid disturbing them further. The scientists measured gill beat rates—how quickly the gill covers are opened and closed—a measure known from earlier studies to be a good indicator of distress in fishes.

All of the trouts were clearly distressed by the treatment they received, but not equally across treatments. In the two control groups, gill beats rose from an original resting rate of about 50 beats per minute (bpm) to about 70 bpm. The gill beat rate rose to about 90 bpm in the bee venom and vinegar groups.

All of the trouts had been trained to swim to a ring to be fed whenever a light went on, but following their respective treatments, none approached the ring, even though they had not been fed for a day. (This contrasts with anecdotal observations of hooked fishes returning to the bait following release.) Instead, they rested on their pectoral and tail fins at the bottom of the tank. Some fishes from the bee and vinegar groups also rocked from side to side, and made occasional darting movements. Some of the vinegar-treated fishes also rubbed their snouts against the tank walls or gravel, as if trying to relieve a sting or an itch.

Toward the end of the first hour, control fishes' gill beats returned to normal. By comparison, gill beats of fishes from the bee venom and vinegar groups were still 70 bpm or more at 2 hours after injection, and they didn't return to normal until 3½ hours later. In addition, at 1 hour post-injection, control fishes began to show alertness when the light came on, although they still did not approach the food ring. One hour and 20 minutes after injection, fishes from both control groups were approaching the food ring and taking pellets as they sank through the water. It took nearly three times as long before the bee venom– and vinegar-treated fishes started showing interest in the food ring.

The trouts' negative reactions to the insults were dramatically reduced by the use of a painkiller, morphine. Morphine belongs to a family of drugs called opioids, and fishes are known to have an opioid-responsive system. Their behavior in response to it here is consistent with their experience of relief of pain by the drug.

In separate experiments being conducted at about the same

time, the ichthyologist Lilia Chervova at Moscow State University was documenting that nociceptors—the nervous tissue sensitive to noxious stimuli—are widely distributed across the bodies of trouts, cods, and carps. She found that the most sensitivity was located around the eyes, nostrils, tail, and the pectoral and dorsal fins—parts of the body that, like our faces and hands, do most of the sensing and manipulation of objects. Chervova also found that the drug tramadol suppressed sensitivity to electric shocks in a dose-dependent manner: more drug, faster pain relief.

The experiments by Braithwaite, Sneddon, and Chervova are strongly suggestive that fishes are *feeling* pain and not merely responding reflexively to a negative stimulus. But there was still another test worth trying, one that would involve a change in complex behavior requiring higher-order cognitive processes. Recognizing and focusing attention on an unfamiliar object seemed just the ticket, and that is what Sneddon, Braithwaite, and Michael Gentle decided to focus on.

Like most fishes, trouts recognize and actively avoid objects that have been newly introduced into their environment. Knowing this, the researchers built a tower of red LEGO blocks and put it into the fishes' home tanks. When they returned the "control" fishes to their home tanks after they had been handled and injected with saline into their lips, these fishes actively avoided the tower, whereas fishes injected with vinegar regularly wandered near the tower. The vinegar appeared to impair trouts' ability to perform a higher-order cognitive behavior—awareness and avoidance of a novel object. The research team conjectured that the pain of vinegar so distracted the afflicted trouts that they were unable to perform normal survival behaviors.

In an attempt to further verify this "distraction" hypothesis, fishes in both treatment groups were injected with morphine following the injection of saline or vinegar. This time, fishes in both treatment groups—saline-then-morphine, or vinegar-then-morphine—avoided the LEGO tower.

Other Studies of Fish Sentience

The experiments I've summarized are not the final word on the matter of fish pain. There are other angles to assessing how fishes respond to what we regard as painful. One of the expectations of a consciously experienced pain as opposed to an unconscious, reflexive reaction to nasty stimuli is a variable or nuanced response. One way to test for this is to vary the intensity of the stimulus. For example, paradise fishes responded to low-level electric shocks by swimming about more actively, as if trying to find an escape route. In contrast, higher-intensity shocks led to retreat from the shock source, and defensive behaviors.

A different approach is to vary the behavioral state of a fish at the time of the stimulus. In a study using 132 zebrafishes, responses to an injection of acetic acid into the tail varied according to whether or not the fishes were frightened before the injection. When injected only, zebrafishes swam erratically and beat their tails in a peculiar manner that did not produce propulsion. However, when they were pre-exposed to the alarm pheromone of another zebrafish, they behaved as zebrafishes normally do when confronted with something new or scary: they either froze in one place or swam near the bottom. They did not swim erratically or beat their tails. The difference suggests that the fishes' fear suppressed or overrode their pain—a phenomenon well known in humans and other mammals. It is an adaptive response because fleeing a dangerous situation that could end in death takes priority over stopping to tend to a wound.

Lynne Sneddon used what I consider to be a most convincing way to examine pain in zebrafishes: she asked if they were willing to pay a cost to get pain relief. Like most captive animals, fishes like stimulation. For instance, zebrafishes prefer to swim in an enriched chamber with vegetation and objects to explore rather than in a barren chamber in the same tank. When Sneddon injected zebrafishes with acetic acid, this preference didn't change;

nor did it change for other zebrafishes injected with saline water (which causes only brief pain). However, if a painkiller was dissolved in the barren, unpreferred chamber of the tank, the fishes injected with the acid chose to swim in the unfavorable, barren chamber. The saline-injected fishes remained in the enriched side of the tank. Thus, zebrafishes will pay a cost in return for gaining some relief from their pain.

When Janicke Nordgreen from the Norwegian School of Veterinary Science and Joseph Garner, now at Stanford University, presented a different method for evaluating pain in goldfishes, it yielded a surprising result. They attached small foil heaters to sixteen goldfishes and slowly increased the temperature. (I was somewhat relieved to read that the apparatus was fitted with sensors and safeguards that shut off the heaters to prevent severe burns.) Half of the goldfishes were injected with morphine; the others received saline. The authors believed that if goldfishes feel the pain of heat, then the morphine-treated fishes would be able to withstand higher temperatures before reacting to it.

Not so. Both groups of fishes showed an appropriate pain response: they began to "wriggle," and it happened at about the same temperature. However, checking on the goldfishes thirty minutes or more after they had been returned to their home tanks, the researchers noticed that those from each group were exhibiting different behaviors. Morphine-treated fishes swam about as they normally would, whereas the saline-treated ones showed more escape responses, including so-called "C-starts" (moving the head and tail toward the same side of the body, forming a "C"), swimming, and tail-flicking (flicking the tail without sideways movements of the head or trunk region).

Garner and Nordgreen's study is evidence that a fish can feel both initial, sharp pain and the lasting pain that follows. The response can be likened to our reaction to putting our hand on a hot stove. First, we have an immediate, reflexive response: we involuntarily jerk our hand away from the heat without pausing to think about it. It is only a second or so later that we feel the true brunt of the pain. Then we may endure hours or days of discomfort while our bodies protect the offended limb and remind us not to do it again! This result suggests to me that goldfishes might have more of those C fibers—the ones associated with lasting, throbbing pain—that trouts were found to have in short supply.

Toward Scientific Consensus

The weight of evidence for fish pain is strong enough today that it has the support of venerable institutions—among them, the American Veterinary Medical Association, whose 2013 Guidelines for the Euthanasia of Animals state:

> Suggestions that finfish [fish that are not shellfish] responses to pain merely represent simple reflexes have been refuted by studies demonstrating forebrain and midbrain electrical activity in response to stimulation and differing with type of nociceptor stimulation. Learning and memory consolidation in trials where finfish are taught to avoid noxious stimuli have moved the issue of finfish cognition and sentience forward to the point where the preponderance of accumulated evidence supports the position that finfish should be accorded the same considerations as terrestrial vertebrates in regard to relief from pain.

In 2012 an august group of scientists met at Cambridge University to discuss the current scientific understanding of animal consciousness. After a day of discussion, a Declaration on Consciousness was drafted and signed. Among its conclusions:

> Neural circuits supporting behavioral/electrophysiological states of attentiveness, sleep and decision making appear to have arisen in evolution as early as the invertebrate radi-

> ation, being evident in insects and cephalopod mollusks (e.g., octopus).

Translation: consciousness needn't require having a backbone. Furthermore:

> The neural substrates of emotions do not appear to be confined to cortical structures. In fact, subcortical neural networks aroused during affective states in humans are also critically important for generating emotional behaviors in animals.

Translation: emotions also derive from parts of the brain outside the cortex. And:

> The absence of a neocortex does not appear to preclude an organism from experiencing affective states.

Translation: you don't need a big, convoluted humanlike brain to feel excited about food or scared of predators.

Now you may be thinking: Bravo, you clever scientists, for coming up with a new way to demonstrate that you are the last ones to recognize what common sense already told us is patently obvious. As the psychologist and author Gay Bradshaw declared: "This is not news, it's Science 101." But it also speaks to the challenge of accepting a phenomenon (consciousness) that is fundamentally private, and to the historical reluctance of science to fully embrace it in anything other than a human.

Fishes show the hallmarks of pain both physiologically and behaviorally. They possess the specialized nerve fibers that mammals and birds use to detect noxious stimuli. They can learn to avoid electric shocks and anglers' hooks. They are cognitively impaired when subjected to nasty insults to their bodies, and this impairment can be reversed if they are provided with pain relief.

Does this close the book on the debate over pain and consciousness in fishes? Not likely. There may always be those who use the crutch of uncertainty to assert that fishes are pain-free. Even if the evidence for the few fish species studied is accepted as true pain, one can still claim that we just don't know for the myriad other fish species fortunate not to have been subjected to scalpels, syringes, or small foil heaters.

Not only is scientific consensus squarely behind consciousness and pain in fishes, consciousness probably evolved first in fishes. Why? Because fishes were the first vertebrates, because they had been evolving for well over 100 million years before the ancestors of today's mammals and birds set foot on land, and because those ancestors would have greatly benefited from having some modicum of wherewithal by the time they started colonizing such dramatically new terrain. Also, it is likely that fishes' ancestors evolved consciousness because fishes today have abilities that are consistent with their being conscious and sentient. As we will discover, fishes use their brains to achieve some quite useful outcomes.

From Stress to Joy

The fish's face is one of its notoriously weak features. Even allowing for the fact that it was the first real face ever attempted, little more can be said for it than that the mouth, nose, eyes, and forehead—if such it can be called—are in the proper order. It is of no use for frowning or smiling; if the fish could do these things, it would receive a great deal more sympathy than it does.

—Brian Curtis, *The Life Story of the Fish*

A woman shared with me a tale of two fishes. In late 2009 she bought a five-gallon tank and three small goldfishes, an oranda, a black moor, and a ryukin/fantail. Like many novice aquarists, Lori knew little about how to care for fishes, and she purchased and lost several goldfishes over the ensuing months. But the original fantail and black moor continued to survive. Lori named the fantail "Seabiscuit" and her husband named the black moor "Blackie."

Lori came home for lunch one day and found, to her horror, Blackie trapped inside a decorative pagoda she had placed in the tank as extra stimulation for her fishes. Struggling to get out, Blackie was bumping repeatedly against the walls and windows of his plastic prison. He looked weak.

Meanwhile, Seabiscuit was darting frantically at Blackie in what Lori took to be an effort to free him from the pagoda. Seabiscuit repeatedly charged at Blackie as though trying to budge him loose. Gingerly, Lori reached for the pagoda, and as gently as possible managed to work Blackie loose with her fingers. He was in poor shape. He had rubbed off all the scales and velvety finish on one side, and his right eye was swollen and raw. He hung listlessly at the bottom of the tank, barely moving. Lori didn't think he would survive.

For the next few days Seabiscuit stayed protectively by Blackie's side, and the little black moor recovered. His eye healed, and a new set of scales gradually grew back on his damaged side.

From that point on, Lori noticed a distinct change in the relationship between Blackie and Seabiscuit, and in her own view of them: "Prior to the pagoda incident, Seabiscuit had been bossy, often chasing Blackie aggressively, but this behavior stopped. I began to perceive fishes as individuals with feelings and personalities."

She moved them into a twenty-gallon tank with a big filter and minimal furnishings. Blackie died in June 2015, age six, apparently due to a faulty filter. Seabiscuit is "hanging in there," with a newer goldfish companion named Too Much, who was rescued from a school carnival.

A separate account, published in a South African newspaper twenty-five years earlier, has uncanny parallels to Lori's. It involved a severely deformed black moor goldfish named—you guessed it—Blackie, who could barely swim. When Blackie was transferred into a tank containing a larger oranda goldfish named Big Red, Big Red took an immediate interest in his disabled tankmate. He also began to provide assistance by placing himself just beneath Blackie. Together they would swim around the tank as a tandem, Big Red providing the propulsion that aided Blackie's mobility and access to food after it was sprinkled on the surface. The pet store owner attributed Big Red's behavior to compassion.

Emotional Hardware

Stories like Lori's and the South African pet store owner's don't carry much scientific weight because they are isolated, anecdotal observations, and behaviors and the emotions that underlie them are notoriously difficult to interpret. For instance, how are we to know that Seabiscuit wasn't attacking Blackie in the pagoda out of fear or stress? For me, the lasting change in the two fishes' relationship afterward is the more telling observation. It suggests that Blackie's mishap was a significant event, and that it brought them closer.

Putting anecdotes aside, what does the science say about fish emotions? A good place to start is the hardware in fishes' brains and bodies.

Emotions involve relatively old brain circuits conserved through evolution and shared by all vertebrates. As we saw in the previous section, you don't need a big brain with a neocortex to feel petrified or pissed off. A growing cadre of experts believes emotions originated alongside consciousness. Reacting is sometimes better than thinking. Imagine you are a early marine creature suddenly confronted by a predator. If you have to think to yourself, "Gee, I better get out of here," you will soon be someone's meal. It is more useful to immediately flee in terror and leave the thinking for later.

Emotions are closely linked to hormones—compounds produced by our glands that affect physiology and behavior. How the brain produces hormonal patterns—the so-called neuroendocrine response—is known to be virtually identical in bony fishes and in mammals. The inference is that these patterns might play out similarly in the conscious, emotional realm—that is, the psychoneuroendocrinology of these two groups might also be similar.

Oxytocin provides an example of these parallels. Also known as the "love drug," oxytocin is associated with bonding, orgasm, labor contractions, nursing, and the feeling of falling in love. Researchers from McMaster University in Hamilton, Canada, have discovered that the fish version of this same hormone, *isotocin*, also regulates behavior in different social settings. When adult male daffodil cichlids were injected with either isotocin or a saline solution, the saline-treated controls showed no discernible behavioral changes. In contrast, isotocin-treated fishes became more emotional. They were more aggressive toward a larger perceived rival when placed in a simulated competition over territory. Surprisingly, mid-ranking cichlids injected with isotocin showed submissive behavior toward other members of their shoal. The authors speculate that the submissive response ensures these highly social fishes—which are cooperative child-rearers—remain a cohesive, more stable group. It may not be love (as far as we know), but it is a nice and friendly response.

Another way to investigate fish emotions is to look for parallels with mammals and birds by subjecting their brains to similar insults and comparing the results. One target of such comparisons is the amygdala—a pair of almond-shaped structures making up part of the brain's ancient limbic system. In mammals, the amygdala helps drive emotional reactions, memory, and decision making. The medial pallium of a fish's brain appears to perform the role of the amygdala. When this region is either disabled (by cutting off its nerve supply) or electrically stimulated, changes in aggression occur that mirror those seen in land-dwelling animals subjected to analogous treatment. Studies on goldfishes have also shown that the medial pallium is involved in an emotional response to a fearful stimulus.

How do fishes show fear? For instance, how do they react when attacked by a predator? They respond as we might expect them to if they are feeling afraid. In addition to breathing faster and releasing alarm pheromones, they show classical behaviors shown by land animals when scared: they may flee, freeze, try to look bigger, or change color. For some time afterward they also stop feeding, and avoid the area where the attack occurred.

Might a fish become more relaxed when exposed to drugs that have an anxiety-melting effect on us? Oxazepam is one such drug—used extensively by human patients for the treatment of anxiety and insomnia, and for the control of alcohol withdrawal symptoms. When researchers led by Jonatan Klaminder at Umeå University in Sweden caught wild Eurasian perches and exposed them to oxazepam, the fishes were more active and showed better chances of survival. Increased activity might seem a surprising response to a drug that relaxes people, but the fishes' response is actually consistent with a relaxing effect: tranquil fishes are less afraid to explore their surroundings. In this state, the treated fishes spent less time clustered with their allies and more time foraging, which might also explain their improved survival rate in a captive setting free of predators.

Being relaxed is all well and good if you're in a safe environment, but fear evolved for good reason: it motivates us to flee and hide from danger. Fishes are capable social learners, readily learning to fear something simply by observing the reactions of others of their kind. For example, naive fathead minnows who were initially unafraid of unfamiliar predators swimming on the other side of a glass barrier soon learned to avoid those predators by watching the fearful reactions of experienced minnows.

Fathead minnows also learn to avoid predators when exposed to schreckstoff from other fathead minnows (recall the fishes' alarm pheromone from our discussion of smell). Do they treat these odor-based clues to lurking danger as seriously as they do visual clues? Apparently not. Scientists from the University of Saskatchewan trained fishes that an unfamiliar odor was "safe" because it never led to negative consequences. In fact, the odor was from pike, a dangerous predator of minnows, but the minnows used in this study were collected from a pond where pikes do not occur, and so were presumed ignorant of pike odor and its implications. A group of control minnows received the same training regimen, only with blank water (no pike odor). On testing day, minnows from both training groups were exposed individually to pike odor paired with either (1) fathead minnow schreckstoff, or (2) a knowledgeable and therefore frightened "model" fathead minnow responding to the risky pike odor. Minnows who had no prior exposure to pike odor responded equivalently to the alarm pheromone or the fright reaction of a model minnow. However, minnows taught to believe that pike odor was "safe" showed little response to the alarm pheromone, whereas they showed characteristic fear behavior (moving and foraging less, taking shelter) in response to their fellow fearful fathead.

So, for a fathead minnow, at least, the sight of fear is more persuasive than the smell of fear. The study also supports the idea that, when it comes to predation risk, minnows trust other minnows more than themselves. It is better to heed a threat that turns out to be benign than to ignore a threat that turns out to be real. Or, as the old saying goes: Better safe than sorry.

Stress Relief

Being able to remove oneself from fearful situations is not only important to survival, it favors long-term health. It is well known from unsettling studies of rats, dogs, monkeys, and other species—and indeed, from human victims of war and other prolonged hardships—that unrelieved stress can lead to all sorts of problems, including anxiety, depression, and lowered immunity.

One of our bodies' responses to stress is to release cortisol. This so-called stress hormone acts to regulate stress, and it performs this function in other vertebrates, including fishes.

A team of scientists from the Max Planck Institute of Neurobiology and the University of California studied genetically manipulated zebrafishes with a cortisol deficit. These fishes suffered from consistently high levels of stress, and they showed signs of depression in behavioral tests. When normal zebrafishes are placed in new surroundings they act withdrawn and swim around hesitantly in the first few minutes. But curiosity soon prevails and they begin to investigate their new tank. In contrast, the mutant fishes showed great difficulty becoming accustomed to their new situation, and they had a particularly strong reaction to being alone: they sank to the bottom of the tank and stayed completely still.

The fishes' behavior returned to normal when either of two drugs—diazepam (Valium), an antianxiety drug, or fluoxetine (Prozac), an antidepressant—was added to the water. Social interactions, consisting of visual interaction with other zebrafishes through the aquarium wall, also helped alleviate depressive behavior in the mutant individuals.

If fishes can be vulnerable to depression and anxiety, might they also take an active role in relieving it? Do fishes seek ways to chill out? A 2011 headline, "Calm Down, Dear, I'll Rub Your Fins," describes such a thing. Surmising that the caresses that reef fishes receive from cleanerfishes might heighten pleasure and relieve stress, a research team led by Marta Soares at the Higher Institute of Applied Psychology (ISPA) in Lisbon designed an experiment to test the idea.

They caught thirty-two striated surgeonfishes from a region of Australia's Great Barrier Reef. Once they were accustomed to captivity, the fishes were randomly assigned to either a stressed or nonstressed group. The unfortunate ones assigned to the stressed group were confined for thirty minutes in a bucket with water just deep enough to cover their bodies. This treatment had the intended effect of significantly raising their blood cortisol—a standard measure of stress. Then, stressed and unstressed fishes were individually placed for two one-hour sessions in a separate tank with a handmade, look-alike model of a cleanerfish. The shape and colors of the model closely mimicked a cleaner wrasse, a reef fish that makes a living by providing a cleaning service to customers such as surgeonfishes. In half of the tanks the model was stationary; in the other half, the model was mechanically rigged to move in a gentle sweeping motion.

Stressed surgeonfishes were drawn to the mobile model like kids to candy. They swam over to the false cleanerfish and leaned their bodies right up against it. But they did this only if it was the one that could stroke them. They averaged fifteen separate visits to the mobile model, compared to zero visits to the stationary model. The strokes from the model also brought stress relief, as measured by cortisol levels that dropped when fishes (from both stressed and unstressed groups) had access to a moving cleanerfish model compared to stationary models. Cortisol also dropped in proportion to the time spent in contact with the moving models.

With the characteristic reserve of a scientist, Marta Soares concluded: "We know that fish experience pain, [so] maybe fish have pleasure, too."

Despite the cutesy tone of the media report about fishes rubbing each other's fins, this is not science lite. It unveils important implications about social living, and having a quality of life. It supports the idea that pleasure motivates fishes to visit cleanerfishes, because the moving models were not removing parasites or anything else, yet the surgeonfishes still went to them repeatedly.

Pleasure evolved to reward "good" behaviors that promote flourishing of the individual and perpetuation of its genes; hence, the good feelings we know that come from eating food, playing, staying comfortable, and having sex. Until recently it was considered unscientific to even speculate on how fishes might be feeling emotionally. For this reason most of the discussion has been restricted to the physiology of so-called reward systems. An elegantly simple scientific definition of a reward is anything for which an animal will work.

In mammals, the dopamine system is a key player in the physiology of reward. When rats play, their brains release large amounts of dopamine and opiates, and when they (or we) are given drugs that block the receptors for these chemicals, they lose their attraction for sweet foods they normally enjoy. Fishes also have a dopamine system. If you give a goldfish a compound that stimulates the release of dopamine from his brain—such as amphetamine or apomorphine—the goldfish engages in rewarding behavior: he wants more of the compound. Goldfishes plied with amphetamine prefer to swim in a chamber treated with amphetamine, whereas goldfishes exposed to pentobarbital, a pleasure-squasher, learn to avoid it. Amphetamine produces a rewarding effect in monkeys, rats, and humans, and this happens by increasing the availability of dopamine receptors in the central reward system. Since the goldfish brain has cells containing dopamine, the same mechanism is thought to be responsible for amphetamine's rewarding effects on goldfishes. Like some mammals, fishes are prone to abusing amphetamine and cocaine, unable to resist them when they are freely available. But in the case of those surgeonfishes sidling up to mobile cleanerfish models to get stroked, there is no addiction—just a fish responding to a desire for a pleasurable, therapeutic massage.*

Games Fishes Play

If you've won a prize, made a basket from the three-point zone, or seen a toddler squealing with delight while being playfully chased by a parent, then you know joy. One type of behavior that is joy-inducing is play. Play is useful, especially to young animals who need to develop physical strength and coordination, and to learn important survival and social skills. Play also has a psychological element: it is fun. Scientists have been exploring animal play for a good while; the German philosopher Karl Groos published *The Play of Animals* in 1898.

Animal play is not easily studied. It is a spontaneous activity and the participants generally need to be feeling relaxed or happy to engage in it. Most observations of animal play are serendipitous.

* I'm also happy to report that the surgeonfishes were returned to their homes on the reef after the experiment.

That is no barrier to Gordon M. Burghardt, an ethologist at the University of Tennessee with a striking physical resemblance to Charles Darwin. In a career spanning nearly six decades and hundreds of scientific papers, Burghardt has not shied away from provocative topics, and that includes animal play where you might not expect to find it—or, what he describes on his website as "play behavior in 'non-playing' taxa."

In 2005 Burghardt published the most comprehensive exploration of animal play to date. The cover of *The Genesis of Animal Play* features a tropical fish, a captive male white-spotted cichlid, pushing a submersible thermometer with his nose. Burghardt and two colleagues, Vladimir Dinets and James B. Murphy, have since published a study of three male white-spotted cichlids interacting with this thermometer—a 4.5-inch glass tube with a weight at the bottom causing it to float vertically. Over the course of twelve sessions, the team recorded more than 1,400 instances of the thermometer being nudged by the three fishes, who were placed individually in the tank for each session.

Each fish had his own style. Fish 1 mainly "attacked" the top of the thermometer, causing it to wobble before returning to a vertical position. Fish 2 also liked to swirl around the thermometer, making contacts as he went. Fish 3 batted the object either from the bottom, the midsection, or the top. His hits were the most intense, causing the thermometer to bob about the tank and sometimes get stuck in a corner. Collisions between thermometer and the glass walls were loud enough to be heard from the adjoining room.

Is it play? According to Burghardt, it is play if:

1. it does not achieve any clear survival purpose, such as mating, feeding, or fighting;
2. it is voluntary, spontaneous, or rewarding;
3. it differs from typical functional behaviors (sexual, territorial, predatory, defensive, foraging) in form, target, or timing;

4. it is repeated but not neurotic; and
5. it takes place only in the absence of stressors, such as hunger, disease, crowding, or predation.

The cichlids' behavior fit all of these criteria. White-spotted cichlids are not predatory, and the attacks they made on the thermometer did not resemble normal feeding behavior. The availability or absence of food had no consistent effect on their cavorting with the thermometer. The possibility of sexual behavior was also ruled out. The cichlids' interactions with the thermometer resembled their quick jabs at rivals, but were more repetitive—rather like a boxer practicing on a bag—and were engaged in only when the fishes were alone, unstressed, and perhaps understimulated.

Given that the observation tank had other objects in it, including sticks, vegetation, and pebbles, why were these fishes especially attracted to the thermometer? The authors surmise that it might have been the reactive quality of an object that bounces back after being knocked, just like those old life-size inflatable clown toys weighted at the bottom to spring back upright whenever you hit them. Ethologists try to assume the animal's own perspective. Burghardt interprets the bouncing back as "a simulated counterattack by an opponent that was never successful."

This is an example of object play. When two individuals interact playfully, biologists call it social play. Here's an example, courtesy of a former animal shelter worker based in Virginia. She had once shared a house with her husband, several cats, and a banded cichlid kept alone in a tank. The fish developed a sport with the cats, who would occasionally tiptoe over the bookshelves to drink from "his" aquarium. The territorial cichlid would lie in wait for the appearance of one of these furry invaders, hiding under the cover of some reeds in the corner of his tank. Experience had taught the cats to peer into the depths for any sign of an ambush, but the fish knew that and stayed quiet as a mouse. Only when the cat's tongue descended did he burst into action, propelling himself up through the reeds like a torpedo, hell-bent on taking a chunk out of that raspy organ. If she sensed the underwater eruption, the cat might get her first lap in before tongue and fish met.

In time, the participants in this cat-and-fish game of wits showed signs that it was a welcome diversion from their quiet indoor lives. No blood was ever drawn on either side, but the cats would sometimes come right back—with a cocked head and sly eyes—to play the game again.

That is not just social play, it is interspecies social play.

A third variation on play is solitary play. In 2006, a German speech therapist named Alexandra Reichle witnessed an example of solitary play during a visit to an art exhibition at the House of Art in Stuttgart. She describes the exhibit, titled *Kunst Lebt* (Art Lives), as a fantastic mixture with hidden treasures from all museums of the country. It included a large aquarium (an exhibit from the State Museum of Natural History in Karlsruhe) of about 130 cubic feet holding an exquisite collection of colorful and exotic fishes.

As a fish lover, Alexandra spent a long time watching the goings-on behind the glass. She soon discovered a small, graceful, almond-shaped fish dressed in plush mauve with yellow and electric-blue highlights. (She later identified it as a purple queen anthias, a native of Asian seas.) This one seemed to have a destination. She would swim in one direction along the bottom, then, on reaching the end of the tank, she swerved upward and swam to the surface. Arriving there, she was met by the current of a water pump, pushing the little traveler like a rocket back to the other side. There, she descended back to the bottom and started her circuit all over. Reichle shared with me, "The funny thing is, I would call myself rather a pessimistic person and the first thing I would think is that this is a stereotypy (a functionless, repetitive, neurotic behavior) due to the confinement. But this little fish actually seemed to have a lot of fun."

I asked her why she thought it was fun. "While most of the other fishes were just swimming about with no particular destination, this one looked so determined to have fun. I wanted to tell the others to follow her and enjoy her wild ride on the man-made current."

It is not an isolated account. Burghardt has watched marine fishes in a very tall columnar aquarium repeatedly "riding" bubbles from an air stone at the bottom of the tank to the top. He thinks that this might be fun for the fishes, as it would be for us.

Jumping for Joy?

If bubble riding is fun for a fish, might they also jump for joy? If you've spent any amount of time boating, fishing, or bird-watching at lakes and rivers, you have most likely seen fishes jumping out of the water. I have seen it many times. The law of averages dictates that it usually happens when I'm looking in another direction, and my eyes arrive at the action just in time to see a splash. Occasionally, I'm lucky enough to see the fish itself, and I've seen foot-long fishes and tiny inch-long ones leaping body lengths clear of the water.

Certainly, fishes will exit water in desperate attempts to escape predators. Dolphins exploit the behavior, forming a circle and catching the panicked fishes in midair. But just as we may sprint from fun or from fear, different emotions might motivate fishes to jump. Mobula rays aren't motivated by fear when they hurl their large bodies (up to a seventeen-foot wingspan and a ton in weight) skyward in leaps of up to ten feet before splashing down with a loud slap. There are ten recognized species of mobula rays and their aerial stunts have earned them the nickname "flying mobulas." They do it in schools of hundreds. Most of their leaps are calculated to land them on their bellies, but sometimes they do a forward somersault, landing on their backs. Males seem to be the initiators, so some speculate that there might be a courtship role. Other scientists think it might be a parasite removal strategy. Whatever its function, I posit that the rays are enjoying themselves.

While kayaking in the crystal waters of Florida's Chassahowitzka National Wildlife Refuge, I watched several schools of fifty or more mullets moving in graceful formation. Mullets are as beautiful as they are common here. Their cream-colored tail margins and rear fins, and the yellow-tinged border between their metallic backs and white bellies were most evident when they leaped from the water, a behavior mullets are known for. Most of the time I saw one or two successive leaps by a fish, but one made a series of seven. Each jump was about a foot clear of the water and two to three feet in length.

There are eighty species of mullets worldwide, and nobody knows for sure why they leap. They usually land on their sides, prompting theories that they are trying to displace skin parasites. Another idea is that they do it to inhale oxygen. The so-called *aerial respiration hypothesis* is supported by the fact that mullets leap more when the water is lower in oxygen, but is undermined by the likelihood that jumping costs more energy than is gained by gulping air.

Might these fishes also be leaping for fun—a sort of fish play? Gordon M. Burghardt published accounts of a dozen types of fishes leaping and somersaulting repeatedly, sometimes over floating objects—sticks, reeds, sunning turtles, even a dead fish!—for no clear reason other than entertainment.

So far, nobody has subjected this intriguing possibility to scientific experiment. Maybe someone ought to catch a few smart fishes, put them in a lush tank with all the amenities (including romantic music and a mechanical cleanerfish model), then give them floating objects to jump over.

Half a Bathing Suit

Let me share with you a little story of a feeling we all know well. It's the feeling we get when we pass an accident scene, are handed a wrapped gift, or overhear an argument in a restaurant. It's what we call curiosity.

A scientist from Alaska told me about an encounter with

curious fishes during a honeymoon swim at a vacant beach in Jamaica. She and her husband were snorkeling along a reef. An excellent swimmer, the husband discovered to his dismay that his bride was unable to dive below the surface. After his efforts to instruct her on how to dive failed, he tried a more drastic ploy:

> With considerable effort, he pulled half of my swimsuit off of me, then swam down and hooked it to a coral branch about fifteen feet below. Surely I would have motivation to retrieve it, he told me with a laugh.
>
> Not a nudist by nature, I was quite upset even though we were apparently alone. I tried repeatedly to dive down to get it, but to no avail. All of this frantic activity had an unexpected effect on the local reef fishes. Rather than retreating, they began to gather around us. Then I noticed that Bob was also affected, um, in a very personal way. He swam over to me, and made every effort to fulfill his manly urges. Alas, my own buoyancy prevented any successful conclusion to such efforts. We were amazed, however, at the reaction of the fish. Tiny little blue fish, angelfish, a rainbow of colors, shapes and sizes of reef life, formed a complete circle around us, facing us, watching. Their bodies and tails quivered, causing them to look like a unified, shimmering mass.

The husband finally took pity on her and retrieved the swimsuit. As the passion of the moment subsided, the fishes lost interest and the circle dispersed. That two humans who were making a fumbling attempt to perform an act that puts us all on the same plane were surrounded by shoals of attentive fishes continues to intrigue her; she still wonders what the fishes were thinking, and whether they were feeling the energy generated by the humans' amorous venture.

Given fishes' sensitivity to sensory cues in their aqueous medium, several theories might explain what made voyeurs out of these fishes. As visually oriented creatures, our first inclination is to assume they were drawn to the young lovers' movements. But perhaps there was something about the electrical field or the body chemistry of the two humans that aroused their curiosity. Then again, maybe it wasn't benign curiosity the fishes were feeling, but unease as they monitored the intentions of a pair of potential predators. That, too, might be viewed as curiosity, especially as these were not familiar intruders.

When a fish takes notice of us, we enter the conscious world of another being. There is something exhilarating about that. For sure, studying fish emotions is a challenging scientific endeavor. But as we have seen, there are techniques to probe fish feelings, and the accumulating evidence indicates a range of emotions in at least some fishes, including fear, stress, playfulness, joy, and curiosity.

Exploring how and what fishes think is less fraught with challenges than trying to study what they feel. As we'll see, the field of fish cognition has a lot to show for it.

PART IV

WHAT A FISH THINKS

Nothing is too wonderful to be true if it is consistent with the laws of nature.

—Michael Faraday

Fins, Scales, and Intelligence

> **Every other animal currently considered stupid and boring has its own amazing secrets. It's just that nobody has been able to discover them yet.**
> —Vladimir Dinets, *Dragon Songs*

Through time, evolution sees to it that animals become highly proficient at what is important to them. We cannot climb as well as a chimpanzee, who has four to five times our upper-body strength. We cannot sprint like a cheetah, or hop like a kangaroo, and a speeding marlin would be at the finish line of a 100-meter race before Michael Phelps came up for his first breath. These animals need to move fast for survival more than we do, and natural selection dictates that faster individuals are more likely to carry their genes for fleetness into the next generation.

The same principle applies to mental abilities. If nature presents a mental problem, and solving it confers a big advantage, then over time creatures may gain the capacity to perform cognitive feats we would otherwise think were beyond their grasp just because they are small, or not closely related to us. The modern scientific field of cognitive ecology recognizes that intelligence is shaped by the survival requirements that an animal must face during its everyday life. Thus, some birds can remember where they buried tens of thousands of nuts and seeds, which allows them to find them during the long winter months; a burrowing rodent can learn a complex underground maze with hundreds of tunnels in just two days; and a crocodile can have the presence of mind to carry sticks on her head and float them just below an area where herons are nesting, then pounce when an unwary bird swoops down to collect nesting material. If you didn't know a reptile could demonstrate planning and tool use, don't feel left out; neither did scientists until this came to public attention in 2015.

What about the mental abilities of fishes? Notwithstanding the liberties taken by filmmakers in popular movies like *The Little Mermaid*, *Finding Nemo*, and its sequel, *Finding Dory*, can fishes really think? Let's have a look at what fishes can do with those brains of theirs.

Here's an example of fish intelligence, courtesy of the frillfin goby, a small fish of intertidal zones of both eastern and western Atlantic shores. When the tide goes out, frillfins like to stay near shore, nestled in warm, isolated tide pools where they may find lots of tasty tidbits. But tide pools are not always safe havens from danger. Predators such as octopuses or herons may come foraging, and it pays to make a hasty exit. But where is a little fish to go? Frillfin gobies deploy an improbable maneuver: they leap to a neighboring pool.

How do they do it without ending up on the rocks, doomed to die in the sun?

With prominent eyes, slightly puffy cheeks looking down on a pouting mouth, a rounded tail, and tan-gray-brown blotchy markings along a three-inch, torpedo-shaped body, the frillfin goby hardly looks like a candidate for the Animal Einstein Olympics. But its brain is an overachiever by any standard. For the little frillfin memorizes the topography of the intertidal zone—fixing in its mind the layout of depressions that will form future pools in the rocks at low tide—while swimming over them at high tide!

This is an example of cognitive mapping. The use of cognitive maps is well known in human navigation and was long thought to be unique to us until discovered in rats in the late 1940s. It has since been documented in many types of animals.

The goby's skill was demonstrated by the biologist Lester Aronson (1911–1996) at the American Museum of Natural History, in New York City. Around the time the rats were wowing us with their cognitive mapping skills, Aronson constructed an artificial reef in his laboratory. He compelled his gobies to jump by poking a predator-mimicking stick into one of his constructed tide pools. Fishes who had had the opportunity to swim over the room at "high tide" were able to leap to safety 97 percent of the time. Naive fishes who'd had no high-tide experience were only successful at about chance level: 15 percent. With just one high-tide learning session the little gobies still remembered their escape route forty days later.

It should be noted that these fishes were almost certainly stressed during these experiments, having been captured from their wild homes and confined in foreign surroundings. Indeed, several died of disease during Aronson's study, which suggests they were not thriving in their captive setting.

In a recapitulation of patterns we see in other studies, individual performance reflected experience in their wild microhabitats. Fishes collected from beach areas that lacked tide pools at low tide did not perform as well as their seasoned comrades—though they still performed much better than chance. A recent study has found that the brains of rock pool–dwelling goby species are different from those of goby species that hide in the sand and don't need to jump to safety: the brains of the jumpers have more gray matter devoted to spatial memory, whereas the sand dwellers have a greater neural investment in visual processing.

Frillfin gobies' ability to make mental maps, allowing them to leap accurately between tide pools, is a textbook example of having a well-honed mental skill wrought of necessity. As the biologist and author Vladimir Dinets, an authority on behavior and cognition in crocodilians, says: "When people use the word 'intelligence' what they usually mean is 'being able to think the same way I do.'" It's a pretty self-centered way to view being smart. I suspect if a frillfin goby could formulate a definition of intelligence, it would include being able to form and remember mental maps.

Remembering the Escape Route

Forming cognitive maps and recalling them weeks later illustrates more than a frillfin goby's prodigious talent for avoiding a leap of faith. It also exposes the human prejudice to underestimate creatures that we don't understand. I do not know what the species did to earn it, but the (gold)fish's legendary "three-second memory" still lurks in popular culture (just try Googling it). I still see an investment company's advertisement at airports that uses the goldfish's putative three-second memory to contrast with the importance of our maintaining business connections. (I also wish to declare, with some humility, that my memory sometimes falls shy of three seconds, as when I forget where I absentmindedly placed my mobile phone or my eyeglasses.)

Being able to remember something is as useful to a fish as to a finch or a ferret. Tony Pitcher, a biology professor at the University of British Columbia, recalls a classroom study in an animal behavior course he taught many years ago. The students were exploring color vision in goldfishes. Each fish was assigned a feeding tube painted with subtly differing hues, and the fishes demonstrated their good color vision. After the study, the goldfishes were returned to an aquarium. The following year, some of these same fishes were combined with a new group of novice fishes for the study. When placed in the study habitat, the veterans quickly reestablished themselves in their former tubes, making it immediately evident that each remembered the exact color and/or location of its tube from one year before.

The study of fish memories is not a new thing. In 1908, Jacob Reighard, a professor of zoology at the University of Michigan, published a study in which he fed dead sardines to predatory snapper fishes. Some of the sardines were dyed red, some not. The snappers didn't mind, and gobbled both types. But when Reighard made the red sardines unpalatable by the gruesome method of sewing stinging medusa tentacles into their mouths, the snappers soon stopped eating the red ones. Notably, the snappers still wouldn't touch red sardines twenty days later. This experiment not only demonstrates a snapper's memory, but also his capacities to feel pain and to learn from it.

My favorite study of fish memory comes from Culum Brown, the biologist with a particular interest in fish cognition. Brown is the coeditor of *Fish Cognition and Behavior*, a book that has helped to propel the current revolution in our thinking about fish thinking.

Brown collected adult crimson-spotted rainbowfishes from a creek in Queensland, Australia, and transported them to his lab. They are named for a kaleidoscope of bright colors arranged in bands of scales along their flanks. Adult rainbowfishes are about two inches long, and Brown guessed these ones were between one and three years old. He placed the fishes in three large tanks, about forty to a tank, and allowed them a month to get used to their surroundings.

On testing day, he removed three males and two females at random from their home tanks and put them in an experimental tank, equipped with a pulley system that allowed a vertical net (the trawl) to be pulled along the length of the tank. The mesh size of the trawl was less than half an inch, allowing the fishes a clear view to the other side without being able to squeeze through its holes. A single, slightly bigger hole measuring three-quarters of an inch across was placed at the trawl's center, providing an escape route when it was dragged from one end of the tank to the other.

The fishes were given fifteen minutes to adjust to their new environment, then the trawl was dragged from one end to the other over a period of thirty seconds, stopping just over an inch from the end. The trawl was then removed and placed back at its starting position. This constituted one "run" of the experiment. Four more runs followed, at two-minute intervals. Five groups of five fishes were tested in 1997, then tested again in 1998.

In the 1997 trials the rainbowfishes panicked during the first run, darting about erratically and tending to cling near the tank edges, apparently not knowing what to do to escape the approaching trawl. Most of them ended up trapped between the glass and the net. Thereafter, their performance improved steadily, and by the fifth trial each shoal of five was escaping through the hole.

When the same fishes were retested eleven months later—having not seen the experimental tank or the trawl in the intervening period—they showed much less panic than they had the previous year. And they found and used the escape hole, on the first run, at about the same rate as they had by the end of the 1997 runs. "It was almost as if they had had no break and had ten runs in a row!" Brown told me.

By the way, eleven months is nearly one-third of a rainbowfish's life span. That's a very long time to remember something that has happened to you only on one occasion.

There are many other examples of fishes showing the remembrance of things long past. These include the studies showing hook shyness by carps for over a year, and paradise fishes who for several months avoided an area where they were attacked by a predator. And there are legions of anecdotes, like the story of Bentley, a captive humphead wrasse. When his usual dinner gong was reintroduced after months of disuse, Bentley raced to the spot where his favorite meal of squid and prawns was served.

Living and Learning

Memory is closely intertwined with learning, for to remember something one must first come to know it. "For almost every feat of learning displayed by a mammal or a bird, one can find a similar example in fishes," writes the fish biologist Stéphan Reebs. If you want to impress someone with your knowledge of esoteric fish jargon, try rattling off these types of learning by fishes: non-associative learning, habituation, sensitization, pseudoconditioning, classical conditioning, operant conditioning, avoidance learning, transfer of control, successive reversal learning, and interactive learning.

You can watch YouTube videos of goldfishes being clicker-trained to swim through hoops and push balls into miniature soccer goals. This is achieved through conditioning, or learning by association. On performing the desired behavior, the fish receives a stimulus, such as a flash of light, immediately followed by a food reward. The fish soon learns to associate both swimming through the hoop and the light flash with the reward. In time, the fish will know to swim through the hoop when it sees the light flash alone, and will hopefully perform the task even when no food is given. It's the same approach used to clicker-train dogs, cats, rabbits, rats, and mice.

(With some humility, we may recognize that the fishes are our captives, and we are the ones in control in experiments like these. Many are not given the enrichment and space they need, and instead spend their days in what often amounts to barren confinement, without the companionship of others of their kind, and with few if any places to hide. If the only way for an animal to get food is to push around a ball, he's likely to do it. If we were in a similar situation, we'd probably do it, too. On the other hand, this is still preferable to the common alternative of captive fishes getting no stimulation other than food and whatever activity they can observe going on outside the glass.)

Aquarium fish owners often report how their pets seem to know when it is feeding time. Simple captive experiments bear this out. For example, Culum Brown and his colleagues fed captive *Brachyrhaphis episcopi* fishes (locally known as "bishops") at one end of their tank in the morning and the other end in the evening. Within about two weeks, the fishes were waiting at the appropriate place and time. Golden shiners and angelfishes take three to four weeks to achieve this so-called *time-place learning*. By comparison, rats take slightly less, about nineteen days, and garden warblers learn slightly more complex tasks involving four locations and four time periods in just eleven days. These numbers are only modestly meaningful, for they assume equal levels of interest in food—the motivator used in learning experiments—over time. In fact, fishes normally eat at much slower rates (about twice a day) than small birds (every few minutes), so it is harder to keep them motivated for learning experiments, and their learning rates may appear artificially slower.

Fishes' ability to learn quickly is being used to improve the poor survivorship of hatchery-reared fishes after they are released into the wild. Growing up in captive confinement—swimming in circles, receiving food pellets on schedule, and having no exposure to dangerous predators—is a vastly different experience from surviving in the wild. Lacking the worldly survival skills of their wild compatriots, only about five percent of some five billion captive salmons released globally each year to build numbers for angling survive to full adulthood. Research shows that animals bred and reared in captivity for many generations can lose their ability to recognize predators, probably because the ability confers no survival benefit for them.

But when the biologists Flávia Mesquita and Robert Young from Pontificia Catholic University of Minas Gerais, Brazil, exposed very young Nile tilapia to a taxidermied piranha (wrapped in clear plastic to eliminate the release of odors into the water), then immediately caught them at the bottom of the tank with an aquarium net, the tilapia quickly associated the unpleasant netting experience with the sight of the predator. After just three trials tilapia were swimming quickly away in all directions. This "scatter effect" confuses predators. After twelve piranha-netting experiences, the formerly naive youngsters had modified their anti-predator response by rising to the surface and remaining motionless. Control fishes who were not netted initially steered clear of the piranha model—a typical avoidance response by fishes toward a new, unfamiliar object—then soon just ignored it. When the trained fishes were retested seventy-five days after their last training session, more than half of them remembered what they had learned.

Like most studies of fish cognition, these were performed on bony fishes. How do elasmobranchs (the sharks and rays) score in learning tasks? As early as the 1960s, nurse sharks had matched wits with mice on a black-and-white discrimination task, with both species performing at 80 percent success after five days. Demian Chapman with the Institute for Ocean Conservation Science has shown with playback experiments that oceanic whitetip sharks have learned to investigate fishing boats when they shut down their motors, because that signals that a fish has been hooked, and there's an opportunity to get it before the fisherman does. These behaviors suggest a being with a mind.

In a study of problem solving by a cartilaginous fish, a team of biologists from Israel, Austria, and the United States presented hard-to-reach food to vermiculate river stingrays, a freshwater species from South America. In the wild, these rays forage on small animals such as clams and worms buried in the sand by uncovering them and sucking them into their mouths.

During training sessions, the five young stingrays soon learned that an eight-inch piece of plastic PVC pipe contained a piece of food, and they successfully accessed the morsel by creating water suction to draw it toward them. One of the two females was successful in all of her trials, perhaps because she appeared to watch the other rays before her first attempts. Within two days, all five rays had mastered the task. They used different strategies. The two females used undulating fin movements to create a current inside the pipe that moved the food toward them. The three males sometimes used this technique, but more often they used their disk-like body as a suction cup or they combined the suction and undulation methods. (It isn't certain whether these gender differences were coincidental or whether they actually reflect a gender difference in foraging styles in this species.)

Next the experimenters upped the ante. They attached a black and a white connection piece to opposite ends of the pipe. The black connection piece had a mesh barrier inside that would block passage of a piece of food, while the white connection piece had no mesh. Each ray was tested over eight sessions, and by the end all were successfully extracting the food from the pipe by working from the white end. Interestingly, all five rays changed their strategies during this phase of the study. The shift was generally from using undulating fin movements or suction to a combination of both. One male also blew jets of water from his mouth into the pipe to force the food out.

These experiments show that stingrays not only learn, but that they can innovate to solve a problem. And they show tool use by using an agent to manipulate an object, in this case using water to retrieve food. Furthermore, moving away from a strongly attractive cue—the smell of food at one end of the tube—and trying the other side is not a trivial thing; it means they have to work against their natural impulse to follow chemical cues. That involves flexibility, cognition, and a sprinkle of determination.

Malleable Minds

You might be thinking that the 20 percent failure by mice and nurse sharks I mentioned a little earlier is still considerable, and that animals should perform at 100 percent if they want to be considered smart. But like other animals, fishes are not interested in their test scores. They don't succeed by robotically adhering to set patterns of living. They are evolved to be flexible and curious, to try new angles, to think outside the box (or the tube). Even highly trained fishes will always explore alternatives; it's a productive way to behave in the real, dynamic world. With the ever-present threats of storms, earthquakes, floods, and, nowadays, human incursions, it pays to be light on one's fins.

That said, I'm not even remotely suggesting that intelligence is uniformly distributed among the diversity of fishes. Inevitably there are smarter and duller individuals. Then there are differences relating to a species' natural history. More challenging habitats demand more mental sharpness from their residents. As we saw in frillfin gobies living in different shore environments, variations in the sizes of different brain regions and in their associated intelligences can be found within a single species.

Here's an example of how ecological challenges can affect intelligence, from K. K. Sheenaja and K. John Thomas at Sacred Heart College, Kerala, India. In the wild, climbing perches occupy both still and moving water habitats. Individuals were collected from two Indian streams (moving habitat) and compared for their ability to learn a maze with fishes collected from two nearby ponds (still habitat). To navigate the maze, they had to find their way through a small door in each of four walls in their tank to get to a food reward at the other end.

Guess who learned the route faster? It was the stream dwellers. They learned the maze in about four trials, compared to six trials for the average pond dweller. When the research team added visual landmarks by placing a small plant next to each door, the pond perches improved their performance to almost the level of the stream perches, who performed no better than they had before. Apparently, the pond dwellers found the landmarks useful, while the stream dwellers ignored them.

Sheenaja and Thomas have an elegant interpretation for these behavior patterns. Streams are more dynamic habitats than ponds because they are constantly subjected to the flow of water, including periodic floods. Stones, plants, and other landmarks are unreliable for learning a travel route because they are constantly changing as the water flows through. The most reliable constant is oneself. Thus, the better raw maze performance by the stream fishes may be attributed to their relying more on "egocentric cues" than visual ones. By contrast, landmarks are more reliable in a relatively stable habitat like a pond, so it pays to become familiar with them. (Incidentally, studies that find population-level differences within a single species are interesting for another reason: they illustrate evolution in action. One can imagine that if these populations don't interbreed for many generations they might eventually diverge to the extent that they are unable to successfully interbreed. That would qualify them as separate species.)

The malleable mind of a fish can be trained to correct unwanted behaviors, and that can be useful in captive situations. Lisa Davis, a zoological manager of behavioral husbandry for Disney Animal Programs, described to me how they corrected a behavioral problem they were having with their cobias. These large, sleek fishes grow to more than six feet and 172 pounds. With a gourmand's appetite, they are prone to becoming overweight in aquariums. The cobias under Davis's care were having this trouble as well. During feeding hours, they were out-competing the other fishes. So Davis and her team taught them to swim to a particular station in their environment where they would be hand-fed pieces of food. This removed them from the competitive environment where other fishes were being fed "buffet style" twenty feet away. The other fishes in the tank got their fair share, and the cobias returned to more normal weight. Win-win. "Even their previously bulging eyes reduced back to normal positions," Davis told me.

Similarly, when aquarium residents need medical attention, collaboration is best. Manta rays and groupers at Ocean Park Hong Kong, the Georgia Aquarium in Atlanta, and Epcot Center in Orlando have all learned through positive reinforcement training to swim into stretchers for transport. Using positive reinforcement to train fishes to participate willingly in their care and feeding makes life more interesting and rewarding for captive fishes, and may help dispel former stereotypes about their intelligence.

To this point we've seen that fishes are not dunces, that they display features of having a mind and a mental life. But what about some of the more celebrated forms of intellect, such as the ability to plan, and to use tools?

Tools, Plans, and Monkey Minds

Knowledge comes, but wisdom lingers. —Alfred, Lord Tennyson

On July 12, 2009, while diving off the Pacific islands of Palau, Giacomo Bernardi witnessed something unusual, and was lucky enough to capture it on film. An orange-dotted tuskfish uncovered a clam buried in the sand by blowing water at it, picked up the mollusk in his mouth, and carried it to a large rock thirty yards away. Then, using several rapid head-flicks and well-timed releases, the fish eventually smashed open the clam against the rock. In the ensuing twenty minutes, the tuskfish ate three clams, using the same sequence of behaviors to open them.

Bernardi, a professor of evolutionary biology at the University of California, Santa Cruz, is thought to be the first scientist to film a fish demonstrating tool use. By any measure it is remarkable behavior from a fish. Tool use was long believed unique to humans, and it is only in the last decade that scientists have begun to appreciate the behavior beyond mammals and birds.

Bernardi's video unveils new gems every time I watch it. I initially failed to notice that the enterprising tuskfish doesn't uncover the clam in a manner we might expect—by blowing jets of water from his mouth. He actually turns away from the target and snaps his gill cover shut, generating a pulse of water the same way that a book creates a puff of air when you close it fast. And it's more than tool use. By using a logical series of flexible behaviors separated in time and space, the tuskfish is a planner. This behavior brings to mind chimpanzees' use of twigs or grass stems to draw termites from their nests. Or Brazilian capuchin monkeys who use heavy stones to smash hard nuts against flat boulders that serve as anvils. Or crows who drop nuts onto busy intersections and then swoop down during a red light to retrieve the fragments that the car wheels have cracked open for them.

Like a seaborne celebrity, the tuskfish draws an aquatic audience. Fishes of several types swim up to watch the sand blowing in action, and others briefly join our hero during his swim to the rock, like reporters hoping for a good quote.

Halfway to his destination, our tuskfish stops to try out a smaller rock lying on the sand. He makes a couple of halfhearted whacks, then heads on his way again, as if he's decided this one's not worth his time. Who can't relate to his misguided attempts and how they reflect the fallibility of a mortal life?

These are impressive cognitive feats for any animal. That they are performed by a fish clearly upsets the still commonly held assumption that fishes are at the dim end of the animal intelligence spectrum. Even if this particular tuskfish were a rare Stephen Hawking among fishes, his behavior would be remarkable.

But what Bernardi saw that day was not exceptional. Scientists have noticed similar behavior in green wrasses, also called blackspot tuskfishes, on Australia's Great Barrier Reef; in yellowhead wrasses off the coast of Florida; and in a sixbar wrasse in an aquarium setting. In the case of the sixbar wrasse, the captive fish was given pellets that were too large to swallow and too hard to break into pieces using only his jaws. The fish carried one of the pellets to a rock in the aquarium tank and smashed it, much as the tuskfish did the clam. The zoologist who observed this, Łukasz Paśko from the University of Wroclaw in Poland, saw the wrasse perform the pellet-smashing behavior on fifteen occasions, and it was only following many weeks of captivity that he had first noticed it. He described the behavior as "remarkably consistent" and "nearly always successful."

Hard-nosed skeptics might point out that this sort of thing isn't *real* tool use because the fishes aren't wielding one object to manipulate another, as we do with an axe splitting a log for firewood, or a chimpanzee does by using a stick to get to the tastiest termites. Paśko himself refers to the wrasses' actions as "tool-like." But this is not to demean the behavior, for as he points out, smashing a clam or a pellet with a separate tool is simply not an option for a fish. For one thing, a fish isn't equipped with grasping limbs. In addition, the viscosity and density of water makes it difficult to generate sufficient momentum with an isolated tool (try smashing a walnut shell underwater by throwing it against a rock). And clasping a tool in his mouth, the fish's only other practical option, is inefficient because fragments of food would float away, only to be snatched up by other hungry swimmers.

Just as the tuskfish uses water as a force for moving sand, the archerfish also uses water as a force—only this time as a hunting projectile. These four-inch-long tropical marksmen—sporting a row of handsome black patches down their silvery sides—mostly inhabit brackish waters of estuaries, mangroves, and streams from India to the Philippines, Australia, and Polynesia. Their eyes are sufficiently wide, large, and mobile to allow binocular vision. They also have an impressive underbite, which they use to create a gun barrel of sorts. By pressing their tongue against a groove in the upper jaw and suddenly compressing the throat and mouth, archerfishes can squirt a sharp jet of water up to ten feet through the air. With an accuracy in some individuals of nearly 100 percent at a distance of three feet, woe betide a beetle or a grasshopper perched on a leaf above the backwaters where these fishes lurk.

The behavior is notably flexible. An archerfish can squirt water in a single shot, or in a machine gun–like fusillade. Targets have included insects, spiders, an infant lizard, bits of raw meat, scientific models of typical prey, and even observers' eyes—along with their lit cigarettes. Archerfishes also load their weapons according to the size of their prey, using more water for larger, heavier targets. Experienced archers may aim just below their prey on a vertical surface to knock it straight down into the water instead of farther away on land.

Using water as a projectile is only one of many foraging options for these fishes. Most of the time they forage underwater as ordinary fishes do. And if a meal is within just a foot of the surface of the water, they may take the more direct route, leaping to snatch it in their mouths.

Archerfishes live in groups, and they have fantastic observational learning skill. Their hunting prowess doesn't come preinstalled, so novices can only make successful shots at speedy targets after a prolonged training period. Researchers studying captive archerfishes at the University of Erlangen-Nuremberg, Germany, found that inexperienced individuals were not able to successfully hit a target even if it was moving as slowly as a half inch per second. But after watching a thousand attempts (successful and unsuccessful) by another archerfish to hit a moving target, the novices were able to make successful shots at rapidly moving targets. The scientists concluded that archerfishes can assume the viewpoint of another archerfish to learn a difficult skill from a distance. Biologists call this *perspective taking*. What an archerfish does might not require the same level of cognition as that shown by a captive chimp who carried a disabled starling up a tree to help launch it back to the air, but it is nevertheless a form of grasping something from the perspective of another.

High-speed video recordings reveal that these fishes use different shooting strategies depending on the speed and location of flying prey. When using what the researchers call the "predictive leading strategy," archerfishes adjust the trajectory of their sharp jets of water to account for the speed of a flying insect—they aim farther ahead of the target if it is moving faster. If the target is flying low (usually less than seven inches above the water), archerfishes often use a different strategy, which the researchers term "turn and shoot." This involves the fish firing while simultaneously rotating his body horizontally to match the lateral movement of the target, causing the jet of water to "track" the target on its airborne path. These fishes would do any quarterback proud.

Archerfishes compensate for the optical distortion produced by the water-to-air transition, and they do this by learning the physical laws governing apparent target size and the fish's relative position to the target. Having a generalizable rule of fin like this enables an archerfish to gauge the absolute sizes of objects from unfamiliar angles and distances. I wonder if archerfishes also practice entomology, visually identifying insects in order to know whether they are tasty, whether they're too big to eat or too small to bother with, or whether they sting.

Most likely, archerfishes have been squirting water jets for at least as long as humans have been throwing stones, and I suspect that wrasses were using rocks to crack clams open long before our ancestors started bashing hot metal against anvils in the Iron Age. But can fishes spontaneously invent tool use, as we can when unexpected conditions require us to improvise? In May 2014, a study highlighted an example of innovative tool use by Atlantic cods being held in captivity for aquaculture research. Each fish wore a colored plastic tag affixed to the back near the dorsal fin, which allowed the researchers to identify them individually. The holding tank had a self-feeder activated by a string with a loop at the end, and the fishes soon learned that they could release a morsel of food by swimming up to the loop, grabbing it in their mouth, and pulling on it.

Apparently by accident, some of the cods discovered that they could activate the feeder by hooking the loop onto their tag and then swimming a short distance away. These clever cods honed their technique through hundreds of "tests"—and it became a finely tuned series of goal-directed, coordinated movements. It also demonstrated true refinement, because the innovators were able to grab the pellet a fraction of a second faster than by using their mouth to get the food. That fishes are routinely expected to interact with a foreign device to feed themselves is impressive enough, but that some devised a new way of using their tags shows a fish's capacity for flexibility and originality.

Tool use by fishes, so far as we know, seems confined to a limited number of fish groups. Culum Brown suggests that wrasses in particular may be the fishes' answer to the primates among mammals and the corvids (crows, ravens, magpies, and jays) among birds in having a greater-than-expected number of examples of tool use. It could just be that living underwater offers fewer opportunities for tool use than living on land. But we do know that the tuskfishes (a member of the wrasse family) and archerfishes are prime examples of evolution's boundless capacity for creative problem solving, and they might turn out to have plenty of company among other fishes.

Might we count tigerfishes among them?

Turning the Tables

For millennia, birds have been diving into the water to catch fishes. Pelicans, ospreys, gannets, terns, and kingfishers are among the more spectacular examples of an army of feathered fish foes. Gannets, who measure over three feet long and can weigh eight pounds, launch themselves downward from a height of 50 to 100 feet and can be going sixty miles per hour when they fold back their wings just before impact and torpedo to depths of 60 feet to grab an unsuspecting fish with their pointed beaks.

Sometimes the tables are turned.

In January 2014 at Schroda Dam, a man-made lake in Limpopo

Province, South Africa, scientists documented on film something that locals had reported seeing before. As a trio of barn swallows skimmed just above the water, a tigerfish leaped up and snatched one of the birds out of midair.

Tigerfishes are oval-shaped, silver-scaled predatory fishes of African freshwaters. There are several species, the largest of which can reach 150 pounds. They are named for horizontal stripes along their sides, and for the rows of large, sharp teeth that line their mouths. They are prized by fishermen as a game species.

The swallow capture wasn't an isolated incident. The research team that published it reported about twenty separate swallow-snatching incidents per day, which represents as many as 300 barn swallows meeting their maker during the fifteen-day survey.

Think about that for a moment. Swallows are known for their speed and agility as they maneuver after insects on the wing. These birds are probably going at least twenty miles per hour when they suddenly become fish food. I have a hard time imagining a fish without any presence of mind having any success at catching a swallow in flight. Without planning, I think a million hopeful random fish leaps and snaps at the air wouldn't yield a feather. Even if the tigerfish waited just below the surface for an approaching bird, then launched straight up from the depths—as great white sharks do to catch seals porpoising across the surface—I am guessing said fish would be doing an air-snap at a swallow long gone. But the grainy footage of a successful catch didn't reveal a vertical leap by the fish. Instead, the bird was ambushed from the rear. In the video of the fish catching the swallow, the fish leaps at great speed from directly behind the bird, and overtakes it in midair before splashing back into the water.

The four ecologists describe two distinct methods of attack being used by the tigerfishes. One involves skimming along the surface immediately behind the swallow, then launching to catch it. The other is a direct upward attack initiated from at least 1.5 feet below the surface. The advantage of the first approach is that the fish need make no adjustment for the surface-image shift due to light refraction at the water surface, which from underwater makes the swallow appear to be behind where it actually is. One disadvantage of this method is that it may compromise the element of surprise. Obviously, at least some of these fishes have learned to compensate for the distortion angle of the water surface, or else they would have no success with the second method.

This behavior raises a host of questions. How long have tigerfishes been doing this? How did it originate? How was it transmitted through the tigerfish population? And why aren't swallows taking evasive action to avoid being caught, such as flying farther above the water?

I decided to ask the lead author of the tigerfish bird predation studies, Gordon O'Brien, a freshwater ecologist from the University of KwaZulu-Natal's School of Life Sciences, in Pietermaritzburg, South Africa: "The tigerfish population in Schroda Dam was only established very recently from the lower reaches of the Limpopo River, in around the late 1990s. So the population there is very 'young,'" replied O'Brien. "Although tigerfish are faring well within most of their range, in South Africa they are declining due to numerous human impacts. As a result, tigerfish have been placed on the South African protected species list, and introductions to man-made habitats are ongoing."

I asked O'Brien how the bird-hunting behavior originated. He explained that from a tigerfish's perspective the dam is very small, and that he believes the population has been forced to adapt or perish. He and his colleagues saw many larger individuals in very poor condition around the period when this behavior was first recorded in 2009.

O'Brien also had quite a bit to say about the ways in which bird hunting is transmitted through tigerfish populations: "This seems to be a learned behavior. Smaller individuals are not as successful and prefer a 'surface chase' approach to ambushing and striking from deeper below the surface, where the individual has to compensate for the light refraction. . . . We know that tigerfish are very opportunistic and are attracted to heightened activity of other individuals—they get into some sort of a feeding frenzy. When the swallows return on their migrations the sight is quite spectacular and I think that it is during this period that the younger [tigerfishes] learn the behavior."

Avivory (the technical term for eating birds) is not unique to tigerfishes. Largemouth basses, pikes, and other predatory fishes have been witnessed on rare occasions leaping up to grab small birds perched on reeds near the surface. Large catfishes were recently filmed catching pigeons who come to drink from the shallows of the River Tarn in southern France; they use the same ambush technique used by orcas to catch sea lions, lunging and beaching themselves temporarily as they try to grab the prey with their mouths.

It isn't likely that these fishes are showing off. They may actually be hunting birds out of desperation. Schroda Dam is a man-made habitat built in 1993, and the tigerfishes were introduced there to help boost their populations, which were dwindling elsewhere in South Africa. Earlier study had shown that the Schroda Dam tigerfishes are spending considerably more time foraging (up to three times more) than other local tigerfishes, possibly due to food scarcity in the lake. The behavior may even put the tigerfishes themselves at risk of predation from African fish eagles, who are common in the area. The pigeon-catching catfishes at the River Tarn might share a parallel plight. Introduced in 1983, they have survived, but pigeons are not normally on a catfish's grocery list, and the fishes may be pursuing the birds due to a documented shortage of their usual prey: smaller fishes and crayfish. If necessity really is the mother of invention, then it applies to fishes, too.

The authors of the discovery at Schroda Dam cite published notes from 1945 and again in 1960, from other locations in South Africa, by biologists who suspected that tigerfishes were catching birds in flight. Maybe one enterprising tigerfish made a lucky strike at an unsuspecting swallow, then honed his or her skill through practice. The behavior could have spread through the population by observational learning, which fishes can be very good at, as archerfishes demonstrate.

However it started, it has the hallmarks of flexible, cognitive behavior: it is opportunistic, since it is unusual behavior for the species; it requires practice to develop, and skill (and no doubt many failed attempts) to execute; it is almost certainly transmitted through observational learning; and different methods are used.

As for why the swallows have not learned to avoid tigerfishes by flying higher above the water, there are several possibilities: (a) the swallows simply aren't aware they're being caught by fishes; (b) the birds benefit energetically by flying just above the surface; and/or (c) that's where most of the insects are. It seems doubtful that the birds haven't detected the danger, for it would be hard not to notice a sizable fish bursting from the water to grab a colleague flying nearby. Maybe capture by a fish is too rare an event, and the benefits of foraging near the surface too great, for the swallows to abandon surface flying.

Fishes Versus Primates

If fishes can innovate and learn to perform exacting, risky maneuvers to catch food, can they also reason their way through a space-time puzzle designed by humans? Imagine you are hungry and I offer you two identical pieces of pizza. I also tell you that the one on the left will be removed in two minutes, while the other one will not be taken away. Which piece will you eat first? Assuming you're hungry enough to eat both pieces, you will almost certainly start with the piece on the left.

Now imagine you are a fish—a cleaner wrasse, in this instance—and you are offered a similar situation: two plates of identical food that differ only in their color. If you start eating from the blue plate, then the red plate is removed; if you choose red first, the blue plate is left where it is and you can have both. Since we can't simply tell a fish that the red plate will be removed first, the fish has to learn it by experience. Elsewhere, similar experiments have been done with three species of brainy primates: eight capuchin monkeys, four orangutans, and four chimpanzees.

Who do you think did better? If you guessed it was one of the apes, no pizza for you. The fishes solved the problem better than any of the primates. Of the six adult cleaner wrasses tested, all six learned to eat from the red plate first. It took them an average of forty-five trials to figure it out. In contrast, only two of the chimpanzees solved the problem in less than one hundred trials (sixty and seventy, respectively). The remaining two chimps, and all of the orangutans and monkeys, failed the test. The test was then revised to help the primates learn, and all of the capuchins and three of the orangs got it within 100 trials. The other two chimps never did.

The researchers—ten scientists in Germany, Switzerland, and the United States—then presented the successful subjects with reversal tests, in which the plates suddenly took on the opposite roles. No one took well to this bit of deviousness. And only the adult cleaner wrasses and the capuchin monkeys switched preferences within the first hundred trials.

Several juvenile cleaner wrasses were also tested, and they performed markedly worse than the adult fishes, indicating that this is a mental skill that must be learned. One of the study authors, Redouan Bshary, even tried the test on his four-year-old daughter. He set up an equivalent "foraging" trial, placing chocolate M&M's on distinctive permanent and temporary plates. After one hundred trials she had not learned to eat from the temporary plate first.

The authors draw a key conclusion: "The sophisticated foraging decisions which cleaner wrasses demonstrate . . . are not easily achieved by other species with larger and more complexly organized brains." But these skills did not come out of the blue (so to speak). The wrasses' shrewd choice of which plate to eat from first resembles decisions these cleanerfishes have to make in the wild during interactions with client reef fish. And the logic of the experiment was deliberately designed to mimic that situation. Brain size be damned, if it's critical to a species' survival then that species will most likely be good at it.

Because cleanerfishes make their living by gleaning tidbits from the bodies of other fishes who have their own agendas, they need to be more attentive to the possibility that that food source might swim away at any moment. Bananas don't do this; transient client fishes do. And cleaners get a lot of practice. Even on a slow day at the office, cleaner wrasses service hundreds of clients. When business is booming, they can have more than 2,000 interactions per day with a great variety of clients—some of them "regulars" who are residents of the reef, others (perhaps other species) "visitors" who are just passing through. Cleaners are able to discriminate between the two, and they start by servicing visiting clients who will swim off and visit another cleaner at another station if not inspected immediately. Regulars will still be around later on. Red plate, blue plate.

If you're like me, you're rather disappointed in the performance of the primates in what to us seems like a fairly straightforward mental challenge. "The apes' unexpected lack of success appeared to be due to frustration with the task," write the authors. It certainly isn't because they are stupid. Great apes are renowned for solving puzzles, some of which they do better than humans can. For instance, chimpanzees far outperform humans in a spatial memory task with numbers randomly scattered on a computer screen. They also have the wits to use Archimedes' principle—which exploits an object's buoyancy—when confronted with a peanut sitting at the bottom of a clear narrow tube. Unable to dislodge the peanut or to reach into the tube, they will retrieve water from a nearby source, carry it in their mouths, and squirt it into the tube until the peanut floats within reach. Some inventive chimps will even urinate into the tube. Orangutans make mental maps of the locations of hundreds of fruiting trees in their forests, as well as the schedules on which they produce fruit. They are also renowned for their escape artistry, being able to pick locks, and they have even tricked keepers into giving up their keys.

But those are different types of skills. It also probably didn't help the primates that they were all born in captivity, where food was routinely provided several times per day and was not taken away. By contrast, the wrasses were wild-caught and had to fend for themselves during their lives.

When fishes outperform primates on a mental task, it is another reminder of how brain size, body size, presence of fur or scales, and evolutionary proximity to humans are wobbly criteria for gauging intelligence. They also illustrate the plurality and contextuality of intelligence, the fact that it is not one general property but rather a suite of abilities that may be expressed along different axes. One of the reasons that the concept of multiple intelligences is so appealing is that it helps explain how one person can be an excellent artist or an accomplished athlete yet do rather poorly at, say, mathematical or logical tasks. It diminishes the importance we have historically placed on "intelligence" as defined by a selection of human abilities that's too narrow even for our own species.

To this point, most of what we have explored has involved fishes acting as individuals. But few fishes live alone; most are social creatures, and their societies reveal new facets of their lives.

PART V

WHO A FISH KNOWS

Friendship isn't about whom you have known the longest . . . it's about who came and never left your side. —Anonymous

Suspended Together

We of alien looks or words must stick together. —C. J. Sansom

Take a cursory glance at fishes swimming around a coral reef, and you might think they are just a higgledy-piggledy assemblage of creatures. Take a closer look and you'll notice structure in who they choose to swim with. During my travels around the world as an ethologist, I've been able to watch fishes in a variety of settings both captive and wild. From Florida to Washington, D.C., to Mexico, I've seen the different ways fishes aggregate and move about. Snorkeling at Biscayne Bay and off Key Largo in southern Florida, I encountered many dozens of fish species. Some, such as the stingray who swam away from me in the shallows of a beach, and a barracuda who hung motionless over a reef, were alone. Most, however, swam with others of their kind. Atlantic needlefishes parked themselves near shore in small groups just beneath the surface. French grunts drifted in compact clusters, swaying with the undulating currents. A band of eighteen midnight parrotfishes ambled casually along the bottom, making audible crunching sounds as they gnawed on coral rocks. Yellowtail snappers were less gregarious, but I never saw one alone. Although mixed-species shoals are common, fishes clearly recognize members of their own kind and generally favor their company.

The who-swims-with-who effect is muted in captive aquaria where there are fewer representatives of each kind. On a visit to the Smithsonian Institution's National Museum of Natural History in Washington, D.C., I lingered in front of a live coral reef display. The tank contained about twenty fish species and a smattering of invertebrates: prawns, sea urchins, sea stars, and anemones. A pair of yellow tangs—solid lemon-yellow fishes shaped like a disk with a pointy mouth, the species represented by Bubbles in *Finding Nemo*—rarely strayed more than two inches apart. A duo of damselfishes took turns making repeated dashes to the surface to gulp air before doubling back in an instant. A second damselfish pair swam calmly nearby, staying within inches of each other, each mirroring the other's movements. There were also two sets of clownfishes, a pair who nestled among the filaments of an anemone near the bottom of the tank, and a trio who swam near the surface. I was watching an organized community of autonomous beings with social lives. Even though captive fishes have no say in who they are stuck with, I admire that they still manage to form harmonious relationships.

Aquariums illustrate what science demonstrates: fishes have social lives. They swim together; they recognize other individuals by sight, smell, voice, and other sensory channels; they choose mates nonrandomly; and they cooperate.

The fundamental social unit for fishes is the shoal or school. A shoal is a group of fishes who have gathered together in an interactive, social way. Shoaling fishes are aware of each other's presence, and they seek to remain in the group, but they swim independently, and individual fishes may be facing in different directions at any given time. A school of fishes is a more disciplined form of shoaling, in which fishes swim in a more orderly fashion, each going the same speed and in the same direction and each spaced a fairly constant distance from the next. A shoal of fishes is likely to be foraging, like the midnight parrotfishes I mentioned earlier, whereas a school is more likely to be in transit. A million sardines migrating along the Adriatic coast are schooling. Schools tend to be larger and longer-lasting than shoals.

While snorkeling with my girlfriend off the west coast of Puerto Rico in April 2015, we experienced a large school of fishes—probably scaled herrings—up close. Gazing at the lovely reef colors a few feet below us, we suddenly found ourselves in a cloud of small, silvery-gray fishes migrating northward along the shore. Each fish was about the size and shape of a metal nail file, and each swam about three inches from the others. Their large eyes bore a slightly worried look, and the doggedness of their swimming—propelled by the constant, rapid flickering of their tails—gave an impression of earnestness. Underwater visibility was poorer than usual due to windy weather, and such was the number and density of this school that nothing else was visible beyond them. We were enshrouded in fishes. I turned and swam with them for a few seconds, and felt the eerie sensation of being in motion yet not moving relative to my surroundings. They seemed completely unperturbed by the presence of these two knobby, bumbling apes in their midst. I caught glimpses of silver flashes to my seaward side—the flanks of larger fishes ambushing them from the depths. A minute passed and the little migrants were gone as abruptly as they appeared, continuing on their northward journey.

Why do fishes form large schools like this? Benefits of schooling and shoaling include ease of movement, predator detection, information sharing, and strength and safety in numbers. Many fishes moving in the same direction produce a current, so school members save energy just as a peloton of cyclists cuts wind resistance. There is some evidence that slime shed from the bodies of migrating fishes reduces drag; studies of schooling Atlantic spadefishes indicate that this effect might increase swimming efficiency by 60 percent. A subsequent study of wild-caught captive Atlantic silversides has thrown some doubt on the drag-reducing hypothesis. When researchers added a synthetic drag reducer to a flow tank in quantities far exceeding the amount of mucus that would be sloughed off a school of ten thousand silversides in natural conditions, they saw no relative decline in tail-beat rates by the fishes swimming in the tank.

In a large school of migrating fishes, schoolmates are surely anonymous. But fish shoals contain familiar shoalmates, and research finds that familiar shoals behave more efficiently than shoals of unfamiliar individuals. Familiar shoals of fathead minnows show tighter cohesion, more dashing behavior, and less freezing. Familiar shoals perform more predator inspections, in which one or two shoal members approach a nearby predatory fish essentially to serve notice that the predator has been spotted and is not likely to make a successful surprise attack.

Even when you're surrounded by colleagues, some positions in a school or shoal are better than others. In experiments at the University of Cambridge, the fish biologist Jens Krause saw no tendency for chubs (a kind of minnow) to position themselves with a shoal of twenty individuals when the shoal was undisturbed. But when Krause introduced the fish alarm substance schreckstoff into the water, the chubs suddenly showed a strong preference for swimming near others of their own size. Bigger chubs swam near the center of the shoal, and smaller ones were relegated to the less protected periphery where predators are more likely to strike. Krause could detect no signs of aggression, but somehow the fishes knew their place.

Position in the shoal is not the only antipredator tactic used by groups of fishes. Just being in a group is likely to reduce predation risk due to the confusion effect. Predatory perches, pikes, and silversides, for example, have lower success in picking off prey fishes from larger schools. It is uncertain how the confusion takes place, but one biologist likens the baffled predator to a child in a candy shop who is so overwhelmed by all the choices that she cannot decide which candy to get.

The visual homogeneity of single-species shoals enhances the confusion effect. In a shoal of minnows, individuals marked with India ink are at higher risk of attack from a pike. No wonder black or white mollies presented with a choice of shoaling with either black or white fishes choose shoalmates who match their own color. Avoiding conspicuousness may be another reason (in addition to avoiding parasites) that fishes prefer to shoal with non-parasitized fishes than with shoals afflicted by many parasites (visible by black spots on their bodies).

Beyond the benefit of sheer numbers, there are more active ways in which the collective actions of a large number of fishes reduces the vulnerability of any one member of the group to capture by an enemy. Fleeing schools will perform the fountain effect, splitting into two clusters that quickly swim around each side of a predator fish and reassemble behind it. If the predator turns around, the maneuver is repeated. The fountain effect exploits the fact that while the predator fish is faster, the prey are more agile, and they can better avoid a predator who is facing away from them. The behavior requires the same sort of rapid attunement to others that can be seen in vast flocks of birds that appear to change direction in an instant (even though there are tiny delays among them).

A spectacular variant of the fountain effect is the flash expansion, in which all the fishes in a school dart away from the center as the predator attacks. The spray of fishes may cover ten to twenty body lengths in just one-fiftieth of a second. Despite the speed of this movement, the prey fishes never collide, leading to speculation that they must somehow know which direction they and others are planning to go before they dart away.

Studies with the banded killifish show that they form different-size groups depending on context. Theories developed by behavioral ecologists predict that larger schools are a better defense against predators, while smaller shoals are better for foraging due to lower competition. That probably explains why killifishes presented with food and alarm cues together formed larger groups than when presented with food only, but smaller groups than in the presence of alarm cues alone.*

Who's Who Among Fishes

Under our superficial gaze, individual fishes in a single-species school can appear so alike that we may rightly wonder if they can tell each other apart. Not only can they do so, but Redouan Bshary, a leader in fish behavioral research from the University of Neuchâtel, Switzerland, is not aware of any study of fish society where researchers *failed* to find individual recognition. Fishes' suite of well-developed senses may act individually or in combination for recognizing other individuals and distinguishing one's own kind from other species. In captivity, European minnows, for instance, can be trained to recognize other fish species by smell alone, even though they probably rely on additional cues in the wild. As we know, fishes also can recognize individuals of species other than their own, such as participants in cleaner-client relationships.

Culum Brown has studied individual recognition in fishes. He was curious to know whether it mattered to a fish whose company it was put in. It did matter. In about ten to twelve days guppies became familiar with new guppies in their midst, and they could learn to recognize at least fifteen fellow guppies. Why would this be useful? One reason is that like wolves, chickens, and chimpanzees, guppies establish social hierarchies, and knowing one's position in society is useful. A smart guppy can know when to take advantage of his higher rank over a lesser shoalmate, and when to avoid punishment for insubordination by higher-ranking individuals.

Furthermore, guppies may use this knowledge from a third-party

* Unfortunately for fishes, the antipredator benefit of schooling backfires with human predation, for which equipment has been developed to detect and then reap almost all fishes in a single school.

perspective: they are more likely to behave aggressively toward the loser of a fight that they have witnessed between two other guppies. At the same time, the fighting males know who is watching them, or at least they know what sex the audience is. If the audience is female, they curb their aggression, presumably because females do not like to mate with aggressive males. But if the witness is a third male, they make no effort to restrain themselves. Dominance hierarchies require individual recognition, and these audience effects infer awareness of relative ranks. For example, the east African freshwater cichlid fish *Astatotilapia burtoni* has been shown in experiments to infer that if fish A is higher ranking than fish B, and fish B is higher ranking than fish C, then fish A must be higher ranking than fish C.

There are other ways of exploiting knowledge of who's who among fishes. Experiments with European minnows show that they recognize and prefer to associate with shoalmates who are poorer competitors for food. Individual fishes plucked from groups that had been foraging together preferentially spent time on the side of a tank with less-efficient foragers than on the side with more efficient foragers. Bluegill sunfishes, and probably many other fishes, make the same kind of discriminations.

A fish recognizing another fish is one thing. But can a fish recognize you? As countless aquarium enthusiasts will attest, fishes can and do learn the identities of the humans who care for them. One example came to me from Rosamonde Cook, an ecologist with the Biological Monitoring Program in Riverside, California:

> While doing a postdoc at Colorado State University from 1996 to 1999, I worked in the Fishery and Wildlife Biology Department. Students had set up a freshwater tank in a hallway near my office, and it held a young smallmouth bass. When the students left for the summer there was nobody left to feed the fish and I volunteered to help. After

> a few weeks, I noticed that the bass would swim eagerly to the glass and up to the surface whenever I approached. I thought perhaps he recognized me. I mentioned this to one of the fishery professors, who assured me that fishes do not recognize individual people.
>
> When fall arrived and the hallways filled with students, I continued to observe the behavior of the bass. I sometimes watched him secretly from down the hall, and I never saw him respond to the presence of others. But any time I came near the tank, he came out to greet me, even when I was ten feet away and surrounded by other people. I cannot explain this fish's behavior other than that he recognized me and could pick me out of a crowd.

Cook told me she later released the bass in a large pond on property owned by the university where no fishing is allowed.

In April 2014 I struck up a conversation with a former U.S. Fish and Wildlife Service employee who was netting some minnows in a backwater of the Potomac River, and collecting them in a bucket of water. The little fishes were destined for a home aquarium where he has kept a largemouth bass for years. "I sometimes feed him feeder goldfish from PetSmart," he said, "but this is cheaper."

Given what I had learned from Rosamonde Cook and countless other fish watchers, I asked him if he thought the bass recognized him as an individual.

"Absolutely. I'm the one who feeds him, and if my wife or daughter are in the room he doesn't stir. But if I enter he swims to the nearest corner of his tank and wags his tail like a puppy dog."

Does science back up a fish's purported ability to recognize a human? Yes, according to a study of archerfishes (those clever water squirters). When presented with two human faces, the archers quickly learned to select the one that was accompanied by a food reward.

Border Patrol

Being able to recognize other individuals is useful in maintaining and defending your particular spot of surf and turf. Territoriality is widespread among fishes, which use a number of techniques to say "Shove off!" to trespassers: spreading fins and gill covers to look bigger, swimming in place with exaggerated movements, making popping noises with the mouth, changing colors, chasing, and—usually as a last resort—biting.

One of the best lectures I've ever attended was delivered years ago at a meeting of the Animal Behavior Society. The topic was like a Just So Story plucked from a Rudyard Kipling collection. Renee Godard's study of hooded warblers changed the way I thought about the mind of a tiny bird. Weighing less than half an ounce, hooded warblers have superb navigation skills. Those who survive their annual migration between the eastern United States and Central America return to the same little patch of forest they occupied the previous year. There the colorful little sprites reestablish their residences through song and active border patrol.

Remarkably, Godard discovered that male hooded warblers recognize their familiar neighbors from year to year. By playing back the individual songs of these neighbors, she found that the occupant birds tolerated neighboring males' calls as long as these emanated from the particular males' known territory locations. However, if she moved the loudspeaker and broadcast the same calls from the other side of the occupant's territory, the occupant freaked out. It's as if your next-door neighbor unexpectedly greeted you from the house across the street.

Recognizing an individual's song after an eight-month hiatus, but associating it with a particular location, is quite remarkable for such a tiny creature. And you might be wondering what it has to do with fishes. Enter the threespot damselfish.

Damselfishes comprise about 250 species of small, colorful fishes found in tropical waters of both the Atlantic and Indo-

Pacific Oceans. They include the clownfishes made famous in *Finding Nemo*. Despite their demure name, damselfishes are renowned for their fearlessness in defending their spot on a reef. Many times while reef diving in Puerto Rico I saw yellowtail damselfishes tearing out of their alcoves to chase away larger fishes who strayed too close.

So, can a threespot damselfish recognize his neighbor as Godard's hooded warblers can? Years before Godard studied warblers, Ronald Thresher was conducting studies to address this question. Thresher was at that time a postdoctoral student in marine science at the University of Miami, and he chose to work with threespot damselfishes living on reefs off the coast of Panama.

He came up with a simple and effective method to compare the reaction of territorial threespot damselfishes to simulated intrusions by other threespot damselfishes. First, he identified territory holders. Then he caught "neighbor" damselfishes who shared a territory boundary with the occupant, and "stranger" damselfishes whose home turf lay at least fifty feet away. Next, he placed the neighbor in a transparent gallon bottle and the stranger in another bottle. Finally, holding a bottle in each hand and starting from the territory of the neighbor fish, he slowly moved the two bottles toward the occupant's territory.

Thresher conducted at least fifteen sets of presentations to different damselfish males, noting the point at which the territory holder began to attack, and whether he attacked the two unwitting intruders differently. He also presented pairs of two different species to the occupant threespots: a closely related dusky damselfish from the same genus, and a less related species, the blue tang surgeonfish.

The response to bottled strangers and bottled neighbors was strikingly different. Territory holders vigorously attacked the strangers, ramming the bottle and trying to bite them through the confounding barrier. In contrast, they virtually ignored the neighbor fish in the adjacent bottle. When presented with pairs of the other damselfish species, or pairs of surgeonfishes, residents did not discriminate between the two.

Companion experiments by Thresher determined that these damselfishes were recognizing their neighbors by size and especially by the subtle differences in their color patterns. All fishes were thoughtfully returned to their home territories, where hopefully they were able to reestablish their well-earned places on the reef.

Nobody has since tested whether damselfishes can remember their neighbors after a long absence the way hooded warblers can. Perhaps they don't need to, since they are not a migratory species. But I wouldn't be surprised.

Like damselfishes, some bumphead parrotfish males are also territorial. Named for their bulbous, bony foreheads, these reef giants grow to five feet and 165 pounds. During territorial disputes, a pair of males swim at each other from a distance of several yards, and their foreheads make a loud *crunch!* when they collide. The behavior resembles the head-butting displays of bighorn sheep, and it serves a similar purpose. Two males square off and butt heads repeatedly, until one of the pair balks and swims away. Although they involve violence and are certainly not risk-free, these ritualized contests generally avert serious injury or death, thereby allowing the victor to retain his turf and the vanquished to venture off to greener pastures. Battle-scarred veterans may sport dents in their bumps, which become white through time as scales and skin are eroded away. Surprisingly, it was not until 2012 that head-butting by bumphead parrotfishes—or for that matter any marine fish—was recorded. Scientists conjecture that a reason for our failure to notice the behavior might be that it is becoming less common. As bumpheads become scarcer due to overfishing, there may be fewer competitors to do battle against.

Personality Is Not Just for Persons

Individual recognition and competitive contests hint at another dimension to be found in societies: personality. Personality is well established in land animals. What about fishes?

Some years ago, I ordered takeout from an Asian restaurant in my neighborhood. As I waited for the food, I took to loitering near the entrance, where there was an aquarium housing three garibaldis. Garibaldis are bright-red fishes native to the Pacific Ocean, about eight inches long, and named after Giuseppe Garibaldi, an Italian military and political figure whose followers often wore a characteristic scarlet or red shirt. The three fishes' permanent home at the restaurant—which featured a faux rocky outcrop, a couple of plastic plants, and a bed of colored stones—was barren and monotonous compared to their native reef habitats, where they can live to fifteen years.

As I watched them over several separate visits, three random red fishes resolved into three discrete individuals—a social unit with patterns of behavior. One of the two slightly larger fishes was always separate; I only ever saw him on one side of the aquarium, away from the rocks at the other end where the other two generally swam, about three feet away. I saw postures and interactions that looked submissive, assertive, and affectionate. Once, the loner and one of the pair sparred near the midline, each making brief lunges at the other. There were nudges and nips, but nothing blatantly violent. On another occasion, one of the pair swam on its side near the bottom while the other made gentle pokes at its body with the mouth. In the wild, male garibaldis clear a nest site for their mates. More than once, I saw a cone-shaped depression in the blue gravel that lined the bottom of the aquarium, so I think these fishes were feeling the urge to nest. Male garibaldis are strongly territorial and will sometimes nip at divers who enter their nesting area. I'm guessing this trio was a mated pair and an extra male. Perhaps the outlier would have done better to become a female.

Garibaldis are among many species of fishes that can switch gender at various times during their life cycle.

In total, I spent barely thirty minutes watching these three fishes, which represented a tiny fraction of the mural of their lives. But in that time, I witnessed something that stayed with me. I realized I was watching not just three random fishes, but three individuals with autonomous, independent lives. They were there for about four years; then one day I returned to the restaurant to find them gone, replaced by several smaller fishes of different species.

Those three garibaldis, for all intents and purposes, were individuals with personalities. It seems to be so with all fishes—whether a humble herring, a sea bream in a Chinese food aquarium, or a reef shark named Grandma. To hear Cristina Zenato talk about Grandma, you know she's describing someone she cares about, with a personality: "She has a soft nature, and a way of approaching me with the desire to be petted and touched. She is usually very keen to come to me. Even when somebody else is down there with food and I am some distance away she will approach me before anybody else. Sometimes when I let her go she quickly turns and comes back into my lap."

Grandma is an elderly Caribbean reef shark, and the favorite of Zenato, an ocean explorer, conservationist, and certified dive instructor. Athletic, ebullient, and fearless, Zenato has spent twenty years diving with sharks at her home base in the Bahamas and around the world. She also hyperrelaxes them with gentle stroking, then removes hooks from their mouths. For Zenato, a shark is a someone, not a something—an individual with preferences, attitudes, and persona.

Cristina named Grandma for her pale color, which she likens to the gray hair of an old lady. They have known each other for five years. Grandma's the biggest of the group of Caribbean reef sharks that regularly visit one of Cristina's diving spots. Based on her size, eight feet from nose to tail, Grandma is about twenty years old.

Zenato's affection for this shark appears to be mutual: "She is rather gentle. She loves to come close to me and allow me to pet her. As mutual trust grows, the bonds I create with these sharks are spectacular."

Grandma disappeared for a week in early 2014. Zenato had noticed that Grandma was pregnant, and guessed she had gone to find a secluded place to give birth. Slow to reproduce, Caribbean reef sharks produce a litter of just five or six live-born babies every two years. As the days passed and Grandma still hadn't shown up, Zenato grew worried. Another week went by, and then she was back, noticeably sleeker, having delivered her babies into the ocean's cradle: "She swam faster. She was wanting food following the labor of birth. I could see it in her body language, her posture."

Their reunion brought sheer happiness.

Being around sharks so long has taught Zenato their independent nature. "There is a relationship with sharks that teaches us the true meaning of 'no strings attached,' no expectations. And it makes the relationship different from our human expectations, but also more beautiful. I care deeply about Grandma. When I see her I smile and she brings me joy. She seems to enjoy our relationship, too."

Zenato is no less enamored of the bony fishes she encounters and sometimes feeds on her dives. In one area where she dives often, she has befriended three black groupers: Peanut, The Whisperer, and Secret Agent. She describes them as extremely intelligent, curious, and attuned to her thoughts.

How does she tell them apart? "It is no harder than distinguishing your math teacher from your mother. Their colors, shapes, body features, and behaviors vary from one to the other."

At nearly five feet, Peanut—whose olive-gray body is cloaked with black blotches and brassy spots—is the largest of the three. A bite sustained when she tried to steal a piece of fish dangling from a shark's mouth left her with a disability that prevents her from changing the color of the right side of her face, so when her body turns pale to signify a relaxed mood, the right side retains a black mask. It's a fish's rendition of *The Phantom of the Opera*.

The other two groupers in this trio are also distinctive in their appearance, with Secret Agent being next in size and The Whisperer the smallest. Zenato considers Secret Agent the prettiest. "Her skin is pristine, unblemished with marks or discolorations, and she has a more slender face."

But even if the groupers were identical in size and coloration, they would be as different as pie from cake. In spite of her handicap, Peanut is the extrovert among the grouper trio. As soon as she sees Cristina with food, she swims straight up to her face. Peanut has learned Cristina's signals for "not your turn to eat" (holding a piece of PVC pipe in her hand) and "your turn to eat" (PVC pipe hidden).

"Even without food she will approach me and nudge my hand to be stroked," says Zenato with a smile. "She loves the feeling of my chain-mail diving suit over her skin."

Secret Agent is named for her habit of keeping just outside Zenato's field of vision, hovering behind and below just to the right or left side of her back. Like Peanut, Secret Agent has learned the difference between shark feeding time and grouper feeding time.

The Whisperer is the shyest of the three. She is always hovering behind Zenato's ears, as if whispering to her: "Drop me a fish, drop me a fish!" But like a feral cat, The Whisperer remains distant, never allowing Cristina to touch her.

"If I turn or move she turns and moves with me, always staying out of sight unless I jerk my head fast around and catch her by surprise."

Creatures like Grandma and The Whisperer defy the common prejudice that sharks are terrorists and bony fishes primitive and dull. Natural selection acts on variation across individuals, and for complex creatures with minds and social lives, personality is an expression of that variation. You don't have to have fur or feathers to have personality; scales and fins will suffice.

Fish Bonding

Because fishes don't have expressive faces, we tend to find them difficult to identify with or feel for. (Consider, though, that dolphins also cannot change their facial expressions, but we do not harbor the same prejudices against them. Perhaps this is because their expression looks happy, or because we know them to be big-brained mammals. Or both.) Yet, there are solid evolutionary bases for the formation of close bonds among fishes, for such roles as mating, parenting, coöperation, and security. And there are numerous personal accounts that attest to social connections that go beyond mere acquaintance.

Sabrina Golmassian kept fishes while doing graduate studies in English in New Mexico. She was naive about aquariums and didn't think fishes had much going on upstairs when she acquired a one-inch-long gold barb. Frankie lived alone in an aquarium with a snail and a frog. He would often poke at these tankmates, but didn't get much reaction and seemed bored. So Sabrina purchased a second gold barb. She named her Zooey. Frankie's behavior changed immediately with the new arrival. His body vibrated with apparent excitement when Zooey was introduced, sending ripples through the water. As Golmassian describes it: "He had an instant and natural love for the new tankmate. It was surprising, considering that he had lived alone for so long. I have since had other fish who are afraid [of] or uninterested in tankmates. But this was love at first sight."

Zooey was at first rather uninterested in Frankie. In time, she warmed to him, and the two barbs began a companionable life in the tank.

One day, while Sabrina was cleaning the tank, Frankie jumped out of the holding vessel and landed in the sink. Zooey began frantically racing around the holding vessel in what appeared a state of anxiety. Sabrina hurriedly scooped up Frankie and returned him to the water, where he was immobile and barely conscious. Zooey sprang into action, nudging at him and pushing him off the bottom as if wishing him back to life. Frankie recovered, though his movement remained sluggish for several days. Zooey appeared to be more active while Frankie regained his swimming and cognitive abilities.

We can do little better than speculate at what Sabrina's two fishes might have been experiencing emotionally. Pronounced changes in the behavior of one fish following a traumatic event in another's life suggests an attunement that goes beyond mere coexistence.

Here's another anecdote relating to the social lives of fishes. Maureen Dawley, senior librarian at Carnegie Mellon University, was resting by a small pond one day at Beechwood Farms Nature Reserve near Pittsburgh, Pennsylvania, when she happened to notice two fishes swimming together near the water's edge. She describes what happened next: "One of the fish was having a hard time staying upright and every few seconds would begin to lean sideways as if on its way to turning belly-up. Each time the fish began to lean, the other fish gently guided its companion upright with the side of its body or with a gentle nudge with its nose. It was the first time I had ever seen a fish practice loving kindness."

This description reminds me of the goldfish we met earlier, who would swim beneath his severely deformed tankmate, Blackie, helping him to the surface for food.

Here's an observation that I am sure will be familiar, for it seems a common phenomenon in aquarium fishes. It came to me from John Peters, an associate professor of economics at Marist College, New York. John kept many fishes during his teens, and his most memorable was an oscar cichlid who lived in a tank in John's bedroom. Oscars are predatory, and the only fishes ever to join Oscar in his tank were the unfortunate goldfishes John fed to Oscar. John became quite attached to the handsome fish, to whom he said "good night" each night using the same words and tone of voice.

Over time, John noticed that Oscar would sleep or rest on the side of the tank nearest John's bed, about three feet away. About a year after getting Oscar, John rearranged some things in his room. To accommodate the new layout, Oscar's tank was moved to a different wall, which put him on the other side of the bed. Within days, Oscar had changed which side of the tank he liked to chill out on. Now, whenever John said good night to his companion, the fish was situated just behind the glass nearest the bed again.

Is it friendship? Maybe, maybe not. Many oscars enjoy being gently petted by their humans. Of course, those humans also feed them, so there is the possibility that the fish is merely hoping for a food reward.

Although oscar cichlids have a life expectancy of eight to twelve years, Oscar didn't make it to three. The goldfishes had their revenge. Oscar fell ill one day, and soon, in John's words, he "went crazy," bumping quite hard into everything in his tank, swimming upside down and knocking into things. When he stopped thrashing around, he was just about dead. John later learned that goldfishes are toxic to oscar cichlids.

Anecdotes like these are often lost to the wind. It's unfortunate because they still have value to me as a scientist. Not only can we be touched by them, but they may reveal aspects of animal behavior that science is not ready (or able) to explore. I would like to see scientists and amateurs alike share their observations with each other. In time, we may witness patterns of behavior that could steer enterprising scientists to investigate those phenomena.

Social Contracts

One hand washes the other. —Seneca

Given creatures with personalities, memories, and the ability to recognize one another as individuals over time, the stage is set for a more sophisticated form of interaction: the long-term social contract. Street-side businesses like barbershops and restaurants that provide on-site services rely on both drop-in customers and loyal patrons to sustain their livelihoods. In a competitive world, it is only by delivering a good product that they are likely to build their client base. Customers will not return after a shoddy shave, and if the food was off, well, there are other places to eat. Occasionally, a business is exposed as a fraud. Punishments are meted out, and reputations damaged.

It isn't much different on the reef.

Consider the cleaner-client symbiosis of fishes. It is one of the most complex and sophisticated social systems of any animals—not just fishes. The system works as follows. One or two cleanerfishes signal that they are open for business. They work at specific locations, and may use swimming postures and bright colors to enhance the signal's visibility (a fish's version of the rotating red-

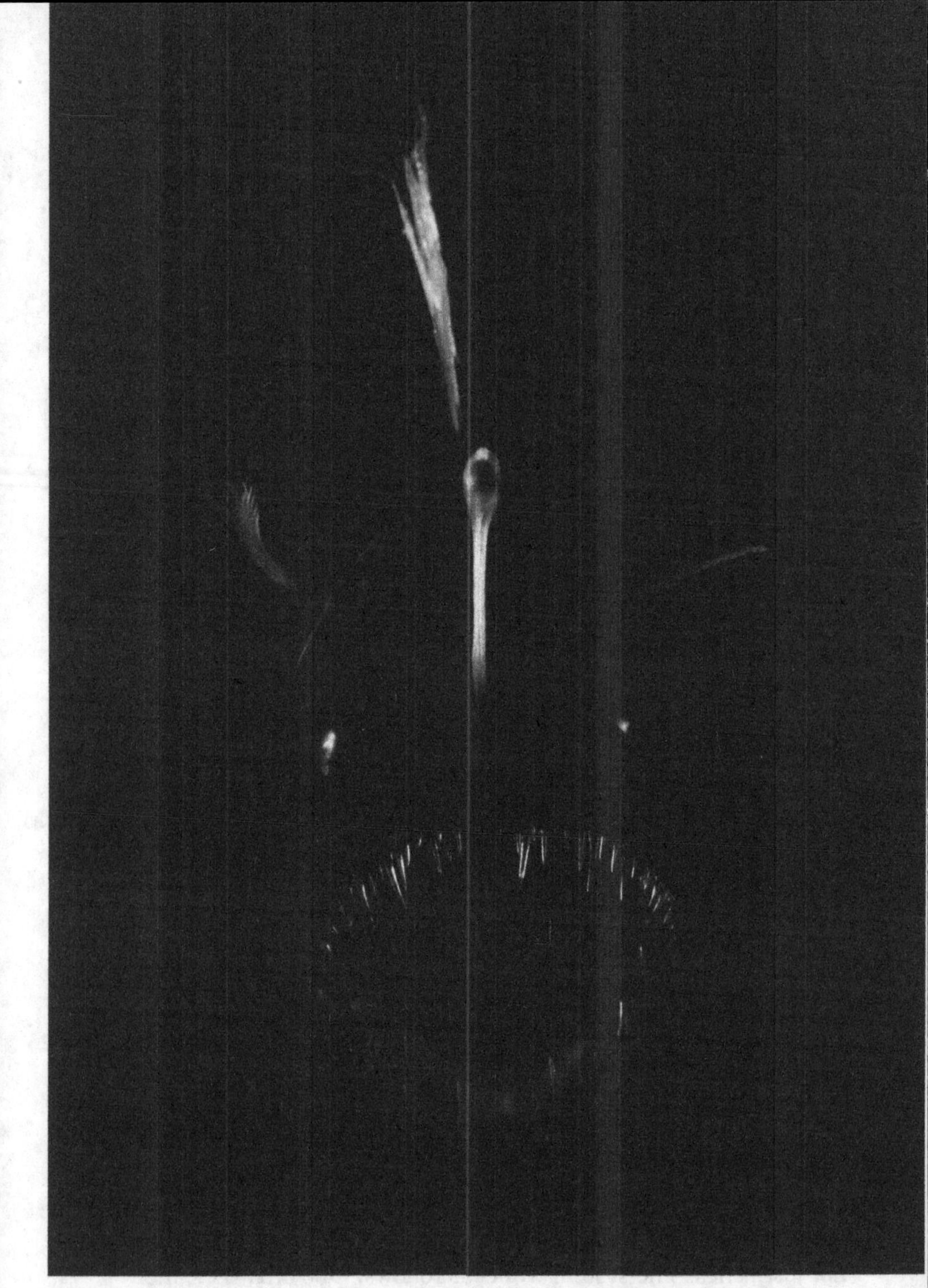

The mysteriousness of fishes is illustrated by this black seadevil. Although they may look like giant monsters from the deep, these anglerfishes rarely reach seven inches in size. The luminescent stalk and lantern act as a lure for unsuspecting prey. (© David Shale / Minden Pictures)

Pleasure motivates useful behaviors. It isn't known for certain why mobula rays leap (here in Oaxaca, Mexico), but some scientists believe they enjoy doing it. (Aaron Goulding Photography)

A pink anemonefish eyes the photographer while retreating into the safety of her chosen sea anemone. (© Mary P. O'Malley)

Lacking hands, fishes are limited in their options for tool use. Here a blackspot tuskfish on the Great Barrier Reef uses a rock as an anvil to bash open a clam. (Scott Gardner)

A small pufferfish spent many hours constructing this circular nest off the southern coast of Japan. He is just visible above and to the left of the center. (© Yogi Okata / Minden Pictures)

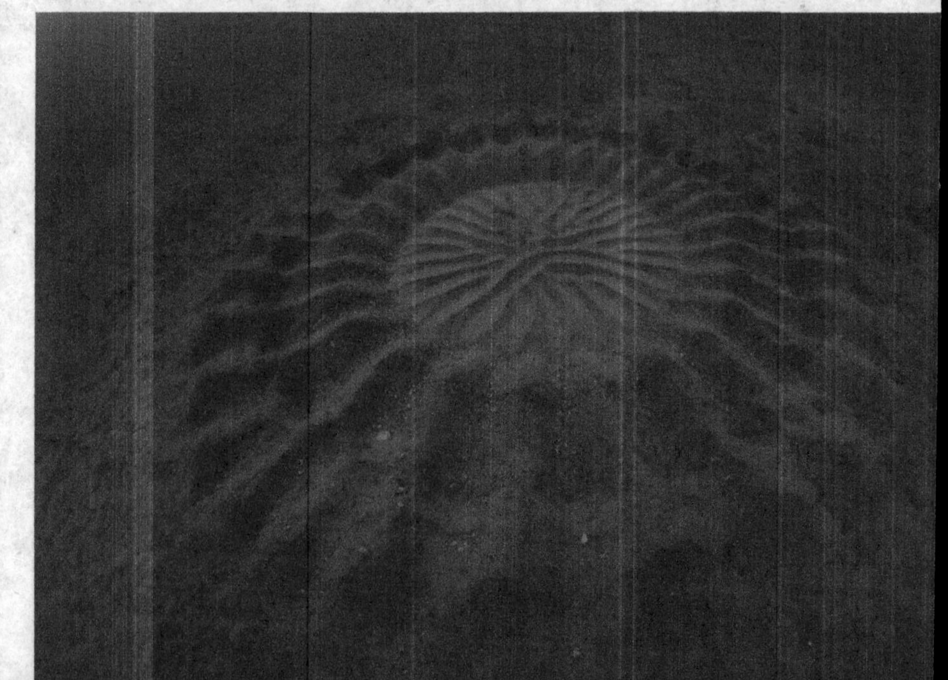

A mother cichlid in Lake Malawi releases her mouthbrooded young after she decides that it is safe to do so. (© Georgette Douwma / Minden Pictures)

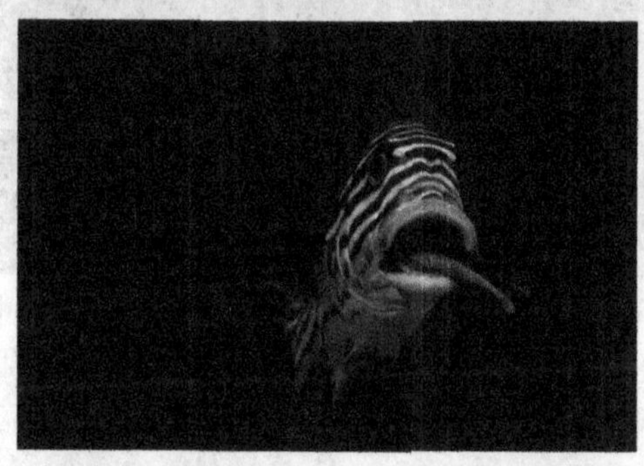

An Indian Ocean oriental sweetlips opens her mouth to allow a blue-streaked cleaner wrasse access for inspection and cleaning. (© Fred Bavendam / Minden Pictures)

Predatory groupers use body shimmies and head-pointing gestures to invite moray eels to hunt cooperatively with them, here in the Mediterranean Sea. (© Reinhard Dirscherl / Minden Pictures)

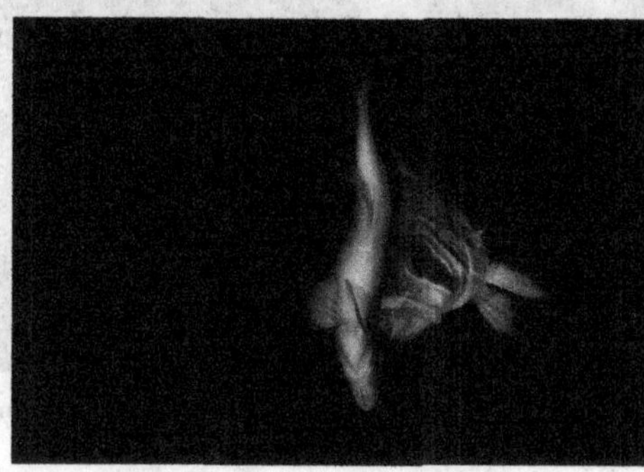

Many fishes court prior to mating. Here a pair of barred hamlets gets in the mood in the Caribbean Sea. (© Alex Mustard / Minden Pictures)

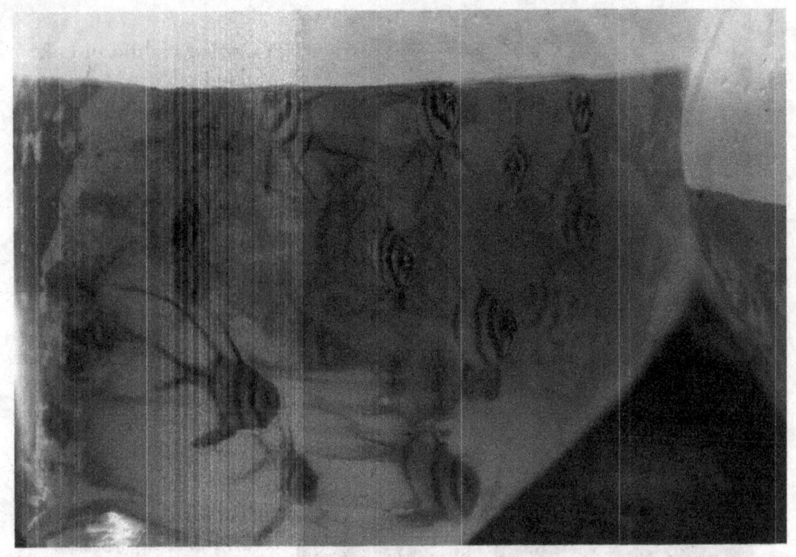

Bangali cardinalfishes, an endangered species caught for the aquarium trade, awaiting shipment to U.S. and European markets from Indonesia. (© Nicolas Cegalerba / Minden Pictures)

A haul of shrimp and bycatch, including many juvenile fishes, on a semi-industrial shrimp dragger, in Mozambique. (© Jeff Rotman / Minden Pictures)

Most fishes, like this purple anthias photographed in open water at Raja Ampat, Indonesia, are able to see a broader color spectrum than we can. (© NPL / Minden Pictures)

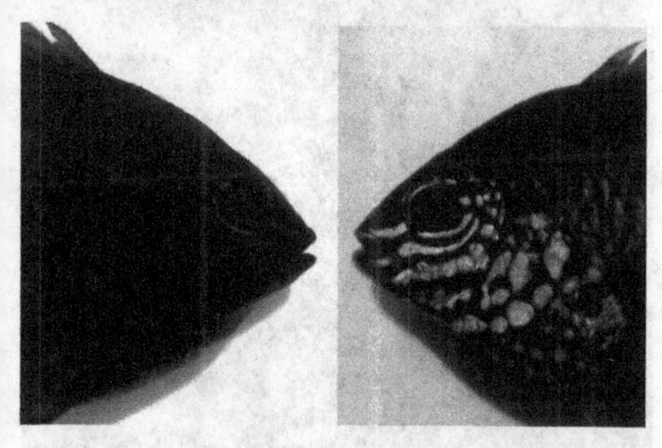

Individual recognition is widespread among fishes. Ambon damselfishes recognize each other by facial patterns visible only in the ultraviolet spectrum. Here are two portraits of the same fish, with the UV image on the right. (Ulrike Siebeck, University of Queensland, Australia)

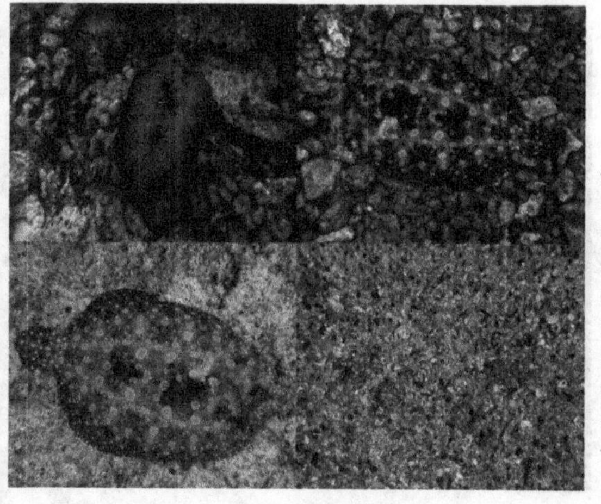

A peacock flounder demonstrates his skill at camouflage. Each of these four images is of the same individual, taken a few minutes apart. In the last frame, the fish is buried in the sand with only the eyes visible.

Tali Ovadia and her nine-year-old fahaka pufferfish, Mango, engaged in one of their prolonged gazing sessions. (Corky Miller)

Cristina Zenato uses gentle stroking to send sharks who have come to trust her (here, three Caribbean reef sharks) into a hyperrelaxed state, enabling her to appreciate them up close, and if necessary to remove fishing hooks from their mouths. (Victor Douieb)

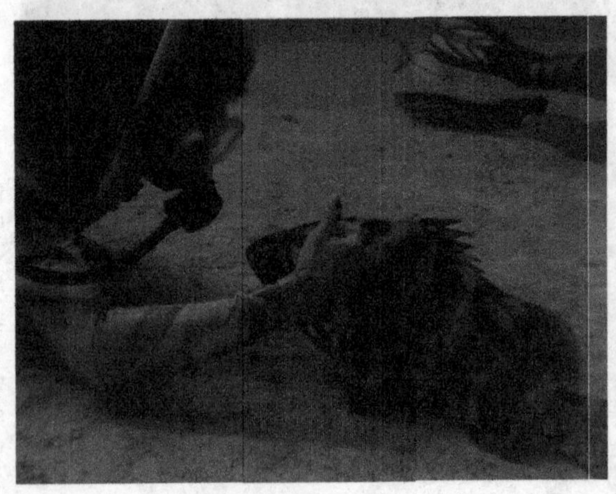

Some fishes grow to trust familiar divers. Here a Bahamian grouper called Larry enjoys a caress from diver Cathy Unruh.

Archerfishes hone their skills through practice and by watching others.
(© Kim Taylor / Minden Pictures)

white-blue cylinder outside a barbershop). Other fishes of various types congregate at the cleaning station, where they wait their turn to be serviced by the cleaners. These so-called client fishes will sometimes signal their readiness for cleaning by assuming a stationary head-up or head-down posture. Cleaners often approach interested clients in a bouncing or tail-wagging manner. They pick over the clients' bodies, removing parasites, dead skin, algae, and other undesirable blemishes. Clients benefit by receiving a spa treatment, including parasite removal. Cleaners get fed.

The diversity of species that ply the cleaning trade attests to its great utility as a profession. Cleaning behavior in fishes has evolved several times independently, and these services have been found in a variety of habitats all over the world. Marine cleanerfishes include many wrasses, some triggerfishes, butterflyfishes, discus fishes, damselfishes, angelfishes, gobies, leatherjackets, pipefishes, sea chubs, surfperches, suckerfishes, jacks, and topsmelts. Freshwater cleanerfishes include cichlids, guppies, carps, sunfishes, killifishes, and sticklebacks. Some invertebrates, including several shrimps, also provide cleaning services. Client lists number well over a hundred known fish species, including sharks and rays. Other clients include lobsters, sea turtles, sea snakes, octopuses, marine iguanas, whales, hippopotamuses, and humans.*

Although I've seen cleanerfishes waiting around for their next client, they can also experience extremely busy workdays. A study on the Great Barrier Reef found that a single cleaner wrasse serviced an average of 2,297 clients each day. Some individual clients visited a specific cleaner up to an average of 144 times per day. That amounts to one visit every five minutes over a twelve-hour daylight period! It sounds like borderline addiction. If the only purpose of visiting cleaners was to have parasites and algae re-

*In some Asian health spas, patrons can pay to dangle their feet in pools where their feet are plucked over by hungry cleanerfishes.

moved, then infestations with these would have to be beyond rampant to justify such a volume of cleanings. This is not to discount the role of parasites in driving cleaner-client mutualisms. Investigations by Alexandra Grutter from the University of Queensland, Australia, found that the average cleaner removes 1,218 parasites from clients per day. When Grutter thwarted access to cleanerfishes by confining one species of client fish—the blackeye thicklip—to cages for twelve hours on the reef, the poor thicklips endured a 4.5-fold increase in parasite loads.

So important is the cleaning station to reef fish communities that cleaners can have a major effect on the diversity of fish species on a reef. When a research team led by Grutter removed cleaner wrasses and kept them out of small reefs for eighteen months at Lizard Island off the eastern coast of Australia, fish diversity was halved and the total number of fishes on the reef reduced to one-quarter for species that move among reefs. The researchers concluded that many fish species—particularly those that migrate between reefs—choose reefs based on the presence of cleanerfishes. This sort of species decline appears to take hold slowly; six months of cleanerfish removals had little impact on diversity.

Clients are not passive participants. When it is their turn, they approach the cleaning station and hover in place, spreading their fins to help cleaners reach all the nooks and crannies. Some open their mouths and gill covers to allow the usually much smaller cleaners to enter and exit. A cleaner will sometimes butt her snout against fins and gill covers to signal the client to spread them for inspection. Cleaners also vibrate their ventral fins so that they tap against the host's body in a signal that says, "Please keep this part of your body still for inspection."

This is a dramatic scene if the client is a large predator. Although a shark or a moray eel could easily snap up the cleaner for a quick snack; it just isn't savvy to eat your service provider.

But it is kosher to show consideration toward them. Groupers, for example, use signals to assist the cleanerfishes who tend to them. A large mouth gape acts as an invitation. While the cleaner is busy, the grouper keeps an eye out for any possible threats. If the cleaner happens to be in the grouper's mouth when danger is approaching, the grouper snaps its mouth closed but leaves just enough room for the cleaner to escape and dart into a safe cranny in the reef. If a cleaner is in the gills, the same thing happens, only this time it is the gill cover that is kept ajar.

Gray reef sharks invite cleaners to service them by angling their bodies upward and opening their mouths wide. The cleaners show no fear as they enter the shark's deadly cavern. They seem to know that this massive predator, hundreds of times their size, means them no harm.

Cleanerfishes have mastered some impressive mental feats, no doubt due to the exacting nature of their craft. The relationship between a cleanerfish and his or her client is not random (recall those 144 visits in one day). It is built on trust, and cultivated over weeks or months. A social contract such as this requires that individual cleaners recognize their clients. With dozens of clients per cleaner, cleanerfishes maintain an impressive mental database of clientele. In choice experiments where a cleaner could choose to swim near one of two clients, the cleaner spent more time near a familiar one. Interestingly, clients showed no such preference in these experimental trials—perhaps because they need only remember the location where the cleaner does business to achieve repeated interactions with the same individual.

In addition to remembering whom they cleaned, cleaner wrasses can also remember *when* they cleaned them. They are more likely to give precedence, say, to a particular triggerfish client who missed their last appointment, because that client will probably have a greater parasite buildup. (It reminds me of hummingbirds' ability to visit specific flowers strategically based on when they last sipped nectar from them.) In experiments in which cleanerfishes were offered foods on plates with four different colors and patterns, the cleaners learned to choose plates that were replenished with food sooner over plates with a slower replenish rate. Cleanerfishes can learn *what* client is the better one to clean. By using memory along three dimensions—who, when, and what—they demonstrate episodic memory, a mental skill held in high regard by biologists.

If a fish can track past events, might she also be able to predict future ones? According to a study done in French Polynesia, roving cleaner wrasses adjust their behavior according to what is called *the shadow of the future*. In humans, this game-theory term refers to our tendency to be more cooperative with a partner when there is a higher likelihood of future interactions. Roving cleaners are more cooperative with their clients near the center of their home ranges, where they are more likely to reencounter client fishes. They do less mucus nipping and cause fewer "jolts" in their clients during cleaning interactions. This study provides us with one of the only examples of a nonhuman animal adjusting levels of cooperation with individual partners to account for future payoffs.

Dubious Dealings

Mucus nipping? Jolts? This is where the cleaner-client symbiosis gets more complicated, even Machiavellian. The cleaner-client symbiosis may seem neat and tidy—where everybody gains, and politeness and consideration for others reign supreme—but a system involving trust and goodwill is vulnerable to exploitation by parties with more selfish interests. As scientists have probed the cleaner-client mutualism more deeply, they have uncovered conflicting interests, and some dastardly deeds.

It turns out that what cleaners most like to glean from their clients is their mucus, which surprisingly has more nutritional value than algae and parasites. Plus it might just taste better. Needless to say, clients do not like having motes of mucus plucked from their bodies. A jolt happens when a client flinches as a cleaner nips at the protective mucus layer that surrounds a fish's body. They may jolt because it hurts, but jolts also inform their cleaners that they bit what they shouldn't have, and that the client knows it.

This conflict of interest between cleaners and clients has a cascade of consequences. Cleaners show extra consideration for clients in the early stages of their relationship. This includes giving them tactile stimulation. They do this by facing away from the client and stroking them with rapid movements of their pelvic and pectoral fins. This caressing behavior seems to be done for two reasons: (1) to encourage a client to stay longer at the cleaning station, and (2) to mollify a client following a jolt. Cleaners are more likely to caress a predaceous client, probably because it lowers the risk of an aggressive chase from a potentially dangerous customer. Hungry predaceous clients receive more caresses than satiated ones, regardless of the client's parasite load. Perhaps there is a real danger of being chased and possibly caught and eaten by an irate client, though I am not aware of any diver having seen it happen.

Predaceous clients are much less aggressive in areas where cleanerfishes ply their trade, so these areas of a reef are considered safe havens. It makes sense that predators would want to be on their best behavior when they are among fishes who provide a valuable service other than as food. I expect also that the tactile stimulation of clients by cleaners has a pacifying effect.

Nevertheless, the great majority of client species are nonpredaceous, which means they lack the predatory client's use of threat to convince a cleanerfish to clean more honestly and with more caresses.

So what's a benign client to do? They have another strategy to keep cleaners on good behavior. It's a sort of tit-for-tat arrangement. Prospective clients watch the performances of cleaners before deciding whether to let a particular cleaner inspect them. By doing this, client fishes accumulate—I'm not making this up—an "image score" for a given cleaner. Think of it as a fish's version of eBay seller ratings. Mucus-nipping cleaners who cause more client jolts are shunned for more honest cleaners. This quality assurance system keeps things more virtuous. Cleaners have reputations, and they pay a price for transgressing into mucus nipping. No wonder cleaners behave more cooperatively with a client when they are being watched.

If a new client, who has no history with the cleaner, is cheated, he or she simply swims away. But a resident client who has built up a relationship of trust with the cleaner behaves as if having been insulted: he chases the cleaner around. Punishment has been shown to make cleaners more cooperative in future interactions.

Cleaning quality also depends on client availability. On reefs where fish clients make fewer visits to cleaning stations, cleaning gobies clean more honestly, ingesting fewer scales relative to the number of ingested parasites. This shift in cleaner behavior toward greater honesty fits a basic tenet of the supply-and-demand economy: when competition for clients is higher, clients have a higher market value, and it pays to give them better service.

The cleaner–client fish mutualism phenomenon represents one of the most complex, well-studied social systems in nature. Redouan Bshary, an authority on this symbiosis, suspects that individual cleaner wrasses can recognize more than 100 individual clients of various species, and remember their last interaction with them. Beyond that, the system encompasses long-term relationships built on trust, crime and punishment, choosiness, audience awareness, reputation, and brownnosing. These social dynamics support a degree of awareness and social sophistication quite out of keeping with our cultural impression of fishes.

While the cleaner-client symbiosis clearly has evolutionary benefits to both cleaners and clients, I maintain that pleasure plays an important role in sustaining it. Pleasure is nature's tool for promoting "good" (adaptive) behavior. Several facets of these interactions support the conclusion that it feels nice. Client fishes actively solicit cleanings, even when they have no parasites or wounds to attend to. And cleaners gratuitously caress clients with their fins to curry favor. Clients also may change color, which probably indicates a change in their emotions signaling a sunnier mood. Feeling pleasure is itself adaptive—witness the therapeutic benefits of massage.

Despite their impressive mental skills, it is doubtful that cleanerfishes reflect on the evolutionary implications of their interactions with clients, or vice versa. I don't see anyone claiming that client fishes visit cleaners because they know it makes them fitter in the Darwinian sense. They do it because they want to.

Keeping Up Appearances

The cleaner-client symbiosis is vulnerable to another, more sinister form of cheating. Different species mimic the cleaners. They look almost identical to them, and they perform all the right moves. Then, when the client least expects it, these little imposters take a bite of fin and dash for cover.

Among the most accomplished imposters are saber-toothed blennies. These little imps are no less crafty than the cleaner wrasses they pretend to be. In one series of experiments, blennies were presented with simulations of client fishes, some of whom retaliated against blenny attacks by chasing the blenny, and some who did not. Retaliation raised the probability that blennies would select different client species, to avoid future attacks. This demonstrates not only that blennies remember past outcomes, but also that the retaliation behavior functions as true punishment. Punishment also serves a "public good" to other members of that client species by steering blennies away.

Conventional evolutionary theory predicts that behaviors should not evolve if other individuals—*free riders*—can benefit from them without any cost to themselves. This prediction raises the question why client fishes expend energy by punishing blennies when the damage has already been done. It turns out that blennies are somehow able to discriminate retaliators from free riders (who don't bother to retaliate but let others of their species do the work), and that free riders are at greater risk of future blenny attacks. So, if you are a client fish who has just suffered the pain of having a notch taken out of your fin by a blenny, it pays to retaliate.

This is a clever analysis, but I find it rather cold and mechanistic. We risk selling animals short when we restrict ourselves to evolutionary calculus. Good heavens, might we also consider that client fishes may retaliate because they have evolved emotions, including one of the most basic: rage? In light of the evidence for emotions in fishes, I feel comfortable interpreting a client's retaliation for the brazen blenny's nip as an angry one.

Cultured

Given its complex social nuances, it would not surprise me if the cleaner-client mutualism involves culture. To biologists, culture is nonheritable information transferred across generations. Human genes don't direct us to wear tattoos or go to the movies, but many people adopt these customs through the example of others. Once thought the sole province of humans, culture is now known to be widespread in mammals and birds, especially in long-lived, social species. Culturally transmitted traits in animals include tool manufacture by crows, migration routes of elephants, the dialects of orcas, and the locations of communal mating sites (leks) in antelopes.

Learning is central to the survival of culture. When I set up speakers and broadcasted recordings of the echolocation calls of foraging bats in fields and woodlands of British Columbia, few bats showed any interest during late spring and early summer. At that time of year only adult bats are flying, and presumably they all know where the best feeding spots are, so why bother attending to a strange new source of bat foraging calls? By August and September, after young bats were weaned and had started flying into the night to forage, it was a different story. My speakers attracted crowds of bats. It seems that young, naive bats were using the calls of older, more experienced ones to find good places to catch insects. Three years later, as I watched millions of Mexican free-tailed bats streaming out of caves at sunset in southern Texas in late summer, I guessed that young ones were also following their older comrades to learn where the good feeding areas are. Nobody was calling it culture then, but when I reflect on the intergenerational loyalty of bats to migration routes, roosts, and feeding sites, it seems appropriate to call it culture.

Do fishes also have culture? Over a twelve-year study of bluehead wrasses on a set of well-studied patch reefs in the San Blas Islands off Panama, Robert Warner of the University of California, Santa Barbara, monitored eighty-seven mating sites of bluehead wrasses. These Caribbean reef fishes remain sexually active year-round, mating on a daily basis. Warner found that long-term mating site selection by these fishes was strikingly constant. The same exact locations remained in daily use over twelve years, which is at least four generations, since this species' maximum life span is only about three years. Furthermore, Warner estimates that there were hundreds of other potential mating sites on these patch reefs that appeared just as attractive, yet for some reason the residents chose not to use them. Also, despite some major population fluctuations during that time, none of the eighty-seven favored love nests fell into disuse. Warner wondered whether these mating sites were favored because they had the best combination of resources. If so, then new fishes ought to use the same sites if the residents were removed.

Warner set about removing the entire local population of bluehead wrasses; then he replaced them with bluehead wrasses from different reefs. Do you think that the newcomers, who soon established mating sites, settled on the sites established by the prior residents? Nope. They established new mating sites, to which they showed just as much loyalty over succeeding generations as did the earlier residents to theirs. In control experiments in which entire bluehead wrasse populations were removed and then placed back on their home reefs, they returned to their former territories (thereby showing that the upheaval of eviction and captivity was not the cause of switching mating sites). Warner concluded that mating site locations are not based on some intrinsic quality of the site, but instead represent culturally transmitted traditions.*

Bluehead wrasses are not the only fishes known to maintain traditional breeding sites according to social convention. Others include herrings, groupers, snappers, surgeonfishes, rabbitfishes, parrotfishes, and mullets. Cultural expression in fishes occurs in other contexts, including daily and seasonal migration routes.

Small fishes have many potential predators, and looking and acting like other group members helps prevent drawing a predator's attention. This might explain the cultural conformity shown by guppies who, having learned a route to a foraging patch by following more knowledgeable fishes, continue to use that route long after the demonstrators are removed. Their chosen path even persists—at least, at first—when a new, more direct route is made available to them. This is quaintly reminiscent of humans who stubbornly cling to a traditional way of doing something even after a newer, more effective method comes along. (Taking handwritten notes comes to mind.) But the guppies only held out for a short while. They soon adopted the efficient option, which goes

*I must admit, I read studies like this with mixed emotions. On the one hand I admire the dedication and creativity of scientists who design and execute ways to test hypotheses. On the other hand, I feel sympathy for the creatures over whose lives we wield so much control. What happened to the residents who were taken from their homes? We may wonder how an animal with culture feels to be removed from his treasured locale.

to show that they are not blind slaves to tradition any more than we are.

Predation by humans presents a sadder side to the loss of cultural knowledge. According to a 2014 study by a team of fisheries biologists and biophysicists, the plundering of fish populations by humans, and our preference for larger (and therefore older) individuals, has disrupted the transmission of knowledge of migration routes among fishes. The researchers developed a mathematical model based on three factors: (1) the strength of the social links between the fishes, (2) the fraction of informed individuals (only older fishes knew migratory routes and destinations), and (3) the preference these informed individuals showed for certain destinations. They found that social cohesion, and the presence of informed individuals, were the most important factors for avoiding loss of coordination and group dissolution.

These cultural disruptions might not be reversible. Culture is not coded into genes, so once lost, it is unlikely to be regained. "If you restore the fish populations, [it] might not be enough," says Giancarlo De Luca, a biophysicist with the study team. "They have essentially lost their group memory." This might explain the failure of many animal populations to recover even when persecution is ended. North Atlantic right whales, western North Pacific gray whales, and many populations of blue whales have shown little signs of growth in the half century or more since we stopped wholesale whaling. The same goes for many species of fishes when their numbers became too low to sustain a commercial catch. As nets and hooks were turned on other species, cod, orange roughy (formerly called by the less palatable-sounding "slimehead"), Patagonian toothfishes (also known as Chilean sea bass), and other long-lived species presumed to have cultural knowledge accumulated over generations, are not bouncing back.

Ocean plundering notwithstanding, as cultured humans we are apt to find virtue in many of our societal activities. Despotic rulers and feudal lords have largely been replaced in the modern era by democracies in which elected leaders are more responsive to the wants and needs of their electorate. Today, resolving regional conflicts is more likely to involve the combined efforts of cooperating nations than in the past. In fish societies, virtue, democracy, and peacekeeping also find their place, as we'll see next.

Cooperation, Democracy, and Peacekeeping

> Nothing truly valuable can be achieved except by the unselfish cooperation of many individuals. —Albert Einstein

In April 2015 I watched some dramatic fish behavior from the second-story balcony of a villa overlooking the Caribbean Sea on the west coast of Puerto Rico. It began with a sudden commotion just offshore about fifty yards along the beach. The surface erupted as dozens of silvery fishes—each about three inches long—leaped en masse from the water. Before they had even hit the water again, more took to the air, reminding me of the finale of a fireworks display. There must have been several hundred in the school. A squadron of larger fins broke the surface at great speed, indicating that predatory fishes were pursuing them.

It was an exhilarating sight. So furious was the energy of the fleeing fishes that my girlfriend and I could hear the sounds of their surging and splashing as they worked along the shore toward us. Time and again, a few seconds of calm were followed by a volley of frenetic activity as sprays of silvery bodies arced through the air, glinting against the evening sun. In their desperation to escape, some fishes became stranded on the beach, flipping and bouncing until the next wave rescued them. A tern swooped down and deftly plucked one from the sand. Others were marooned briefly on a seaweed-covered rock protruding from the shallows.

As the seething school came within yards of our observation perch, we noticed the tight phalanx of the larger fishes—each about one and a half feet long—swimming in parallel just beyond them. Their close alignment, and the effect they were having on their quarry, reminded me of the cooperative hunting performed by dolphins, who encircle a school of fishes or force them ashore, and snatch the less lucky ones as the prey make desperate leaps to safety. There was no encircling in this instance, but the team of hunters seemed to be using the shoreline to corner their prey before ambushing them.

What we saw from the balcony bore little resemblance to the popular cartoon image of a small fish about to be eaten by a larger fish, which is about to be eaten by a still larger fish, ad infinitum. For me, that cliché characterizes a fish as little more than an automaton blindly responding to the hunger urge. What we witnessed was cooperative hunting by fishes. It wasn't a first. Cooperative hunting is known to occur in several fish species. For example, shoals of barracuda will swim in a tight spiral, herding prey into shallow waters for easier hunting. Similarly, the parabolic shape of a shoal of hunting tunas indicates that tunas hunt cooperatively.

Lions are renowned for their cooperative hunting prowess. So are orcas. Scientists don't know how lions signal to each other that it's time to go hunting, but the lions obviously do.

Is it possible that fishes signal their intention to hunt?

A good place to start is the lion's marine namesake, the lionfish. They are named for their "mane" of long, ribbonlike, poisonous pectoral fins, but in hindsight they could as soon have been named for their cooperative hunting style. A 2014 study of two species of lionfish describes a distinctive flared-fin display used to signal to another the desire to hunt together. The soliciting fish approaches the other with his head down and pectoral fins flared, then rapidly undulates his tail fin for a few seconds, followed by slow waving of alternate pectoral fins. The receiving fish almost always responds with fin waving, then the pair moves off to hunt. Hunting in this study involved the cooperating pair cornering a smaller fish using their long pectoral fins and then taking turns striking at the victim. The display looks the same in both species of lionfish, and cooperators are sometimes mixed-species pairs. Cooperators had higher success rates than lionfishes hunting alone. They took turns, sharing the spoils with their hunting partner. It makes sense that they would share, because selfishness would soon erode the desire to cooperate.

The hunting style of yellow-saddle goatfishes further resembles lion-style hunting by assigning different roles to team members. These streamlined, foot-long reef dwellers—usually dressed in yellow but changeable to pinks and blues—hunt in coordinated groups that include chasers and blockers. Chasers flush the prey out of their hiding crevices, and blockers prevent their escape. By coordinating different, complementary roles, goatfishes achieve a quite sophisticated, collaborative method of hunting.

Fish-hunting alliances can be more elaborate still. Groupers and moray eels combine the tactics of lionfishes and goatfishes, using signals or gestures to communicate their desires or intentions, and adopting complementary roles to catch prey. The first description of this behavior came in 2006, when Redouan Bshary and three colleagues described roving coral groupers from the Red Sea using a rapid, full-body shimmy to recruit giant moray eels to hunt with them. The two teammates swim off over the reef like friends on a stroll. The researchers saw dozens of these interactions, and they were able to show that groupers and morays who worked as a team caught more prey fishes than either hunter working alone. The reason for this success is the complementary role each fish plays. The eel is able to pursue fishes into narrow spaces in the reef, whereas the grouper is more effective in the open water surrounding the corals. The poor victim finds itself out of escape options.

The most impressive aspect of the grouper-moray signaling is one of the least obvious: it occurs in the absence of the physical manifestation of the ultimate goal. There is no prey present when a grouper signals to a moray her intention to go hunting. The grouper (and probably the moray) is anticipating and creating a future event. This represents another example of planning. Commenting on the collaboration, the primatologist Frans de Waal wondered if there really is anything that a fish cannot do, adding: "If it comes to survival, highly intelligent solutions are within the reach of animals as different from us as fish."

In 2013 another research team uncovered a variation of cooperative communication and hunting by Red Sea groupers and their partners, only this time the signal resembled something we do to communicate the location of a hidden object: pointing. Roving coral groupers and their close relative leopard coral groupers use a "headstand" signal to indicate the location of hidden prey to cooperative hunting partners of several types: giant moray eels, humphead wrasses, and big blue octopuses. Although the context is similar, this gesture is fundamentally different from the full-body shimmy, because the headstand actually points to a fish or some other edible creature that has just hidden out of reach of the grouper. That makes it a referential gesture, which, outside of humans, had only previously been attributed to great apes and ravens—two groups known to be Einsteins of the animal world.

The headstand signal meets the five criteria for referential gestures, as proposed by the biologists Simone Pika and Thomas Bugnyar based on studies of communication in ravens:

1. It is directed toward an object (prey hiding in a reef crevice);
2. It is purely communicative and mechanically ineffective (i.e., it doesn't directly catch prey);

3. It is directed toward a potential recipient (e.g., a moray, a wrasse, or an octopus);
4. It elicits a voluntary response (e.g., a moray comes over and seeks the prey); and
5. It demonstrates intentionality.

This is neat stuff. Finger-pointing is regarded as a critical communication and social skill, and a crucial milestone in child development. When a child points at something, she is initiating joint attention—that is, she wants you to engage with the item of interest she is pointing at.

The groupers were very patient, sometimes waiting for 10, and as many as 25 minutes over the location. Sometimes, a hunting partner (e.g., a moray eel) was too far away to notice the grouper's pointing gesture. In those cases, the researchers witnessed the grouper swimming over to the moray and performing the body shimmy. The invitation often worked: the moray swam with the grouper back to the crevice where the prey was hidden.

In a follow-up captive study, the same team concluded that the collaborative capacities of groupers compare favorably with those of chimpanzees. Groupers were given a choice to hunt with either of two realistic fake moray eels (life-size photographs laminated in clear plastic and operated by hidden cables and pulleys). One of the fakes was an effective collaborator that flushed out prey, whereas the other swam in the opposite direction. On the first day of testing, the groupers showed no preference for either of the fake eels. But by the second day they had identified the successful partner and preferred it at a rate that compared favorably to that of chimpanzees. When food was out of reach, and getting it required collaboration, the groupers also were at least as proficient as chimps at determining when they needed to recruit a collaborator, doing so in 83 percent of cases. The fishes learned more effectively than chimps when a collaborator was not necessary.

Does this imply that groupers are as smart as chimpanzees? No. How does one even compare the two, when the ape lives on land and has grasping hands, while neither applies to the fish? What it does show is that when needs dictate, a fish is capable of performing smart, flexible behaviors. Alexander Vail thinks you could view the cooperative hunts of coral groupers and morays as a kind of social tool use: "A chimp can get a stick and probe honey out of a hole. A grouper has no hands and can't pick up a stick. But it can use intentional communication to manipulate the behavior of a different species with the attribute it needs." The clever science writer Ed Yong summed it all up with a piece titled "When Your Prey's in a Hole and You Don't Have a Pole, Use a Moray."

Democracy

For me, the beauty of the grouper hunting collaborations is their conscious intentionality. It is two fish minds working in concert to translate and transfer desires and intentions into favorable outcomes.

Another way that desires may lead to social outcomes is through collective decision making. "One common property we see in animal groups from schooling fish to flocking birds to primate groups is that they effectively vote to decide where to go and what to do," says Iain Couzin, an evolutionary biologist at Princeton University. "When one fish heads toward a potential source of food, the other fish vote with their fins on whether to follow," he adds. And this highly democratic process helps animals make decisions as a group that are better than those of any single member.

A benefit of consensus decision making is that speed and accuracy of decisions increase with group size, because they efficiently combine the diverse information possessed by group members. For example, misinformed golden shiners are less prone to committing an error when they are swimming in a group. It appears that groups either aggregate the information into a quorum response, or they follow a few informed experts or leaders.

Decisions on whom to follow may also be informed by the appearance of individual fishes. All else being equal, a healthier, more robust fish knows how to take care of himself, and may thus be considered a better decision maker than a frailer individual. Might a fish make such a discrimination? To investigate, a collaborative team of biologists based in Sweden, the U.K., the United States, and Australia designed experiments in which stickleback fishes were introduced into the center of a Plexiglas tank that contained two identical refuges of attractive rocks and vegetation situated in opposite ends of the tank. Near the rear wall of the tank, starting in the center, a pair of plastic models of sticklebacks "swam" toward opposite refuges by being towed at constant speed along a monofilament wire. One of the models in a pair was made to look healthier than the other. For example, a larger model looks fitter than its smaller counterpart because it appears to be a better forager and a longer-term survivor; a plumper model with a distended abdomen also appears better fed; whereas a model with dark spots looks like it might be harboring parasites.

The sticklebacks behaved as if they had previewed the study plan. When just one fish was presented with two models, it followed the healthier-looking model to its refuge about 60 percent of the time. Performance steadily improved with group size to over 80 percent in groups of ten sticklebacks. This is an example of consensus decision making.

A more sophisticated tool has been developed for studying fish democracy. "Robofish," a realistic, swimming fish robot to which fishes such as minnows respond naturally, is helping scientists gain further insights into the value of collective behavior. Whereas singleton sticklebacks are susceptible to a robofish leader behaving in a maladaptive way (going toward a predator), the quorum response of a larger shoal usually avoids this pitfall. If enough fishes defect, the remainder are more likely to follow the defectors.

Similarly, small shoals of mosquitofishes will follow a robofish toward an arm of an experimental Y-maze in which a predator waits, whereas larger shoals are more likely to shun the robot's lead and instead choose the safer arm of the Y-maze.

A word about realistic fakes, models, and replicas: just because the fishes respond to them doesn't mean that they perceive them as real. Also, keep in mind that the fishes are in artificial conditions and alien surroundings. For this reason they often need to be acclimated to captivity over weeks or even months before they will be calm enough to behave "normally." Acutely perceptive fishes may recognize a man-made model as somehow abnormal, but the motivation to avoid a fearful stimulus may override their doubt.

Peacekeeping

Predatory encounters are not the only dangers animals face. Fishes have to contend with conflicts among their own kind, but because injury and death are bad outcomes when you need to survive and reproduce, actual physical fighting between rivals is rare. Like other animals, fishes often use ritualized displays of strength and virility to avoid more serious physical conflict that would risk injury to either or both opponents. Fishes deploy a range of tactics to impress upon others that combat is not a good idea. They make themselves look as big as possible by spreading their fins, opening their gill covers, or performing a lateral display to show their full size. Booming sounds accentuate size and strength, while water-displacing tail beats add physical force to these saber-rattling displays. Other embellishments include head shakes, body twists, showing off brightly colored body parts, or performing color changes.

Not all displays are intended to show aggression. Fishes also appease. An effective appeasement display involves exposing vulnerable body parts, a tactic that boosts the authenticity of the gesture, as when wolves expose the throat, or monkeys the genitals. The aggressively territorial blunthead cichlid appeases another with a quiver display involving the presentation of the luminous yellow band around the fish's vulnerable midriff.

If troubles nevertheless escalate, cichlids may act as peacemakers. This happens in the golden mbuna, another Malawian cichlid. In captive settings these butter-colored beauties with white-trimmed black racing stripes along their sides form a linear dominance hierarchy, and most interactions occur between dominance-rank neighbors. A male will actively intervene in female-female disputes. He breaks up the altercation without favoring either combatant. If one of the females is unfamiliar, she is favored, and his intervention has been shown to increase the chances that the new female will settle in the group. That, of course, is a most welcome addition for the male, who gains a new potential mate.

Animal hierarchies commonly line up according to body size, with larger individuals ranking higher. As with the alpha male elk who gathers a harem and seeks to keep other males from mating with "his" females, in some fish societies only the largest male gets to mate with an available female. A subordinate male risks having to fight with a larger male if he is within about 5 percent of the bigger one's size. A loss could bump him down a few notches in the mating queue. What is a little fish to do? In an admirable show of restraint, male gobies of various species deliberately limit their food intake to retain their place in the queue.

But dieting gobies are not wearing halos; restraint might yield long-term benefits. In their social groups, which number a dozen or so, a goby normally moves up the community ladder only when a superior dies. Since dieting has been shown to improve longevity in many animals, fasting might also be a good long-term strategy to become a breeder.

In a social setting, restraint and cordiality are the defaults for even the most pugnacious of creatures. One day, pity compelled Lori Cook of Tampa, Florida, to rescue some Siamese fighting fishes (also familiarly known by their genus, *Betta*) isolated in cups at a local Walmart store. She took good care of them, setting them up in a small background pond. As her local reputation as a fish friend grew, so did her collection of bettas, as unwanted fishes from the neighborhood were delivered to her. She also eventually acquired some females from a PetSmart at one dollar each; female bettas have little interest for most pet shoppers because they are not aggressive, and we perceive them as drab.

Lori's experience with these fishes defies their combative reputation. Every morning, she would go out to feed them and they would gather at the pond edge for food. Since bettas are tropical and even in southern Florida temperatures might get too low for them, she uses an aquarium heater during the colder months. Despite having now kept many generations of Siamese fighting fishes, many of whom are males, Lori observes: "I have never, ever, seen two males fighting, nor have I ever seen any evidence of fighting, such as bites or mutilated fins."

Why are these supposed warrior fishes so passive? Probably because getting along is better than fighting. One of the problems with pitting male bettas against each other is the artificiality of captivity, which thwarts the natural urge for a "loser" to flee. The subordinate's efforts to defuse the situation by removing himself are blocked, creating the possible impression for the dominant fish that his rival has changed his mind and wishes to fight again. This, I suspect, is why contests staged in a tank can, reportedly, end in death.

The mental powers of bettas aid them in avoiding dangerous fights. Studies by Rui Oliveira and his colleagues at the ISPA Instituto Universitário, in Lisbon, find that rival males monitor other males' performances in disputes, and they show greater deference to known winners than to losers. Males who had watched other male contests were less eager to approach and display at males whom they had seen winning compared to males they had seen losing, whereas there was no such discrimination between *unseen* winners and losers.

Deception

With all this restraint, cooperation, and peacemaking going on in fishdom, you could be excused for thinking every fish is an "angelfish." Not so fast. As we saw with the cleaner-client symbiosis, any form of cooperation or social interaction opens up opportunities to manipulate for personal gain. Like humans, fishes deploy a dizzying array of visual and behavioral deceptions to fool others. It isn't a big leap from "fish" to "selfish."

Some deceptions are simple ploys to evade detection by predators. During their most vulnerable period as small juveniles, many fishes mimic other creatures that advertise their toxicity with gaudy colors. The shape and color of a juvenile pinnate batfish is strikingly similar to that of a toxic flatworm, while black spots on a pearly white background render a young barramundi cod into a doppelgänger for another toxic flatworm species.

Behavioral embellishments can magnify the deception. In 2011, Godehard Kopp of the University of Göttingen, Germany, captured on film a superb example of mimicry by a fish off the coast of Indonesia. As Kopp filmed a mimic octopus—itself a masterful imitator of other marine organisms—creeping across the sand on a foraging foray, he noticed (barely) a black-marble jawfish flickering among the octopus's tentacles. The fish bore the exact coloration and markings of the cephalopod, and enhanced her camouflage by orienting her body parallel to the invertebrate's arms. Scientists reporting the association speculated that the deception allows the jawfish—which otherwise spends most of its adult life near the safety of a sand burrow—to venture farther away from its home den to forage in relative safety. This is one of the only known examples of a mimic mimicking a mimic.

The tools of mimicry and camouflage are not only available for avoiding predation, they are also used by predators to sneak up on their prey. In freshwaters of both South America and Africa, leaf fishes have evolved to mimic dead and decaying leaves that float or sink to the bottom. Through a combination of visual and behavioral trickery, these patient hunters catch small fishes who stray too close. Leaf fishes position themselves strategically, floating or dangling to blend in optimally with the available foliage. Their tiny, transparent pectoral fins work in overdrive to hold them in position. A ragged, fleshy protrusion from the chin looks just like a decaying petiole, a tasty morsel to an unsuspecting goby. Once the little fish is within range, the predator engulfs it in the vacuum created by its extensible jaws. It's over in less than a quarter second.

In a morbid variation on the leaf fish's deception, some cichlids of the *Nimbochromis* genus in East Africa's Lake Malawi feign death by lying limply on their sides on the lake bottom. When an inquisitive scavenger fish comes to investigate, the "corpse" springs to life, seizing and consuming the curious investigator.

Trumpetfishes and pipefishes combine two popular children's games to sneak up on their prey: they play hide-and-seek by riding piggyback on parrotfishes. The small fishes they are hoping to catch are unfazed by the vegetarian parrotfish, and they often fail to notice the pipefish, who slides off to attack any who come within range. Trumpetfishes will also join a passing school of smaller fishes, avoiding easy detection by possible prey as they lurk within the crowd. This cunning is impressive in itself, but I find no less fascinating the tolerance shown by the accomplices they hide among, who appear to be unafraid of this predator in their midst.

Living in the perpetually dark oceanic abyss, deep-sea anglerfishes do not need to hide. But they are famous for their own brand of deception, in the form of dorsal fins that function as convincing fishing lures. Chances are you've heard of anglerfishes, whose bizarre shapes and gaping mouths remind me of the gargoyles that decorate medieval church facades. Perhaps you didn't know that it is only females whose dorsal fin nature has wrought into a filamentous stalk, referred to by specialists as the *illicium* (whose

Latin root means "to entice" or "mislead"), on the end of which resides the luminescent lure, or *esca*. Deep-sea anglerfishes are a diverse group numbering 160 known species, and their lures come in an array of temptations that would challenge any fisherman's tackle box. Some lures resemble a worm, which the anglerfish causes to squirm appealingly by twitching muscles at the base of the stalk. In species that live in shallow waters, these lures are brightly colored. Those of deep-sea anglerfishes, who live beyond the light's penetration, trade color for light, which is produced by bioluminescent bacteria living in special compartments built into the filament. In some species the tip of the esca bears a lens, which transforms the adjustable filament into an elaborate, tubular light guide—nature's answer to fiber optics. In another species the lure is wriggled within the anglerfish's mouth, sealing the fate of any little fish who ventures unsuspectingly inside (or any quite big fish, for that matter, since anglerfishes can engulf prey their own size).

Is an anglerfish aware of the deception she is creating when she wriggles her dorsal fin lure? It is one of the challenges of any question relating to animals' mental lives. The skeptic can claim that the fish has no awareness of what it is doing, pointing to insects that use mimicry to fool birds and other would-be predators. While I do not wish to disparage insects, anglerfishes, leaf fishes, and trumpetfishes are not members of that invertebrate group. They are full members of the vertebrate league, with brains, senses, biochemistry, and minds befitting that status. Making a living as a fish in the inky depths requires considerable resourcefulness and know-how, especially when your prey are other vertebrates with minds of their own.

To this point I've explored how fishes perceive their worlds, how they feel both physically and emotionally, their thoughts, and their social lives. The main conclusion we may draw from these aspects of what a fish knows is that fishes are individuals with minds and memories, able to plan, capable of recognizing others, equipped with instincts and able to learn from experience. In some cases, fishes have culture. As we've seen, fishes also show virtue through cooperative relationships both within and between species.

There is one vital aspect of a fish's social life that I have yet to explore, and it is the ultimate aim of all organisms: to make more of themselves. When the time is right, the urge to reproduce rivals that most basic need, to find food. True to their diversity, fishes have devised an ocean's worth of methods to procreate and to parent.

PART VI

HOW A FISH BREEDS

"How do you spell 'love'?" —Piglet
"You don't spell it . . . you feel it." —Pooh

—A. A. Milne

Sex Lives

Fishes . . . are characterized by a level of sexual plasticity and flexibility that is unrivalled among other vertebrates.

—Thavamani J. Pandian, *Sexuality in Fishes*

True to their magnificent diversity of forms, fishes exhibit a Full Monty of breeding systems—thirty-two in all. They have as many different kinds of reproductive behaviors and strategies as exist in all the other vertebrates combined.* There are promiscuous fishes, polygamous fishes, and monogamous ones, including fishes that mate for life. Depending on his sexual playbook, a male fish may keep a harem, defend a territory, spawn in a group, engage in sneak copulations, bide his time as a satellite male, or commit acts of sexual piracy. And as we'll see, females are not passive accessories.

The great majority of fishes exhibit a familiar pattern with an obscure name, *gonochorism*, in which individuals are either male

*Thavamani Pandian credits Japanese researchers as being the most energetic in bringing to light the sexual habits of fishes. Some scientists have spent more than five hundred hours underwater to collect data for a single study. Advances in SCUBA technology have greatly helped advance our understanding of fish sex.

or female throughout life. But you can guess what that implies: there are scores of fishes who cross gender lines. For some reason, reef living in particular has had a diversifying effect on sexual expression. More than one-quarter of all fishes on a reef can transition from male to female, or vice versa—with no need for expensive surgery. Other fishes opt for a unisex approach, assuming both male and female identities simultaneously, or sequentially.

Of the species that produce both sperm and eggs at the same time (*simultaneous hermaphrodites* for jargon lovers), most are found in the vast darkness of the deep sea. Being able to fertilize yourself is a very useful adaptation where the daily prospects of finding another of your kind are almost as dim as your surroundings. The sex-changing fishes (*sequential hermaphrodites*) are not so restricted, and they benefit by being different sexes at different ages and sizes. For example, in a mating system in which one male may monopolize many females, it pays to start out as a female, then become a male when you are large and more physically capable of rebuffing the challenges of competitors. Often, all younger members of a species are females, with a harem male occupying the top-dog position. In other cases, the chain of command is reversed, with a string of lesser males awaiting the future prospect of becoming a breeding female.

The popular clownfishes of *Finding Nemo* fame rely on size, hierarchy, and sex change to maintain social order. They live in groups of two large and several smaller individuals. The big ones are the breeding pair, the larger of whom is the breeding female. The subordinates, all males, are ranked hierarchically according to size. Although these lower-ranking fishes may be as old as the spawning pair, the behavioral dominance of the sexually mature individuals keeps the subordinates from growing or developing. Hans and Simone Fricke, who studied this strict mating system, described the low-ranking males as being, in essence, psychophysiologically castrated. Each retains his place in the queue until there is a job opening in the executive office. If the breeding female dies, the chief male changes sex to female and the next largest fish in the subordinate group bumps up to chief male. So there is always hope for a suppressed male in a clownfish family. (All of this reveals a slight inaccuracy in the course of events in *Finding Nemo.* The fact is, upon Nemo's losing his mother, his dad, Marlin, should have become his new mother.)

Sex-changing fishes behave appropriately, performing male-typical or female-typical sexual behaviors according to their current gender assignment. Sexual behavioral plasticity can also be observed in fishes that do not normally change sex, but who are subjected to hormonal manipulation. Although how this happens is not clear in a fish, results of field and laboratory observations suggest that some bony fishes (as distinct from the cartilaginous sharks and rays) have a sexually bipotential brain that can manage two types of behaviors, unlike most other vertebrates, which have a discrete sex differentiation of their brain and can only perform gender-typical sexual behavior.

The ability of individual fishes to change sex shows just how fluid gender divisions can be in nature. If you are at all tuned in to societal trends, you'll know that gender lines are becoming more blurred in humans, too. The book *Becoming Nicole,* for example, explores the social challenges faced by a human family whose son, an identical twin, sought gender reassignment at an early age. As medical advances expand our options to assert our true gender identities, we unwittingly become more like a fish.

Seducing with Artistry

Once you know what sex you are, there is still the matter of whom to mate with. It is not a trivial decision. A sex partner is going to be the one who supplies the other half of the genes that go into your kids, so you want them to be of good quality. It helps to have ways of evaluating the caliber and desirability of a prospective mate. That's where courtship comes in. We date, dine, go to dances, share gifts, and use other methods to test prenuptial waters. So, in their own ways, do many fishes, who seduce prospective partners with dance sequences, love songs, and sensual touches.

And, in at least one kind of fish, with art. We do not typically think of fishes as artists, at least not beyond the sort of passive artistry to be found in the beautiful patterns and colors many fishes display on their bodies. But art is what surprised the veteran Japanese diver and photographer Yoji Ookata during a dive off the southern tip of Japan. At a depth of about eighty feet lay a six-foot-wide symmetrical circular pattern in the sand. The mural featured two concentric rings of ripples radiating outward from a center disk. It was as though a five-hundred-foot-tall giant had waded into the ocean and pressed his thumbprint into the sand.

Mystified as to what might have created this exquisite curiosity, Ookata returned some days later with a film crew, and the mystery was soon solved. The geometric "crop circles" were created by a small, quite ordinary-looking male pufferfish. Swimming on his side while fluttering a pectoral fin, the five-inch puffer spends hours making his masterpiece. He inspects it as he goes, decorating his mural with bits of small shells that he cracks in his mouth before sprinkling them into the central grooves.

Mandalas made by other males have been found since. No two are the same. There appear to be several functions of these structures. Chiefly, they attract female puffers who, all going well, lay their eggs in the inner circle. The furrows help prevent eggs from being carried away by currents, and the crushed shells probably enhance this effect while providing camouflage for the eggs. Males who build more elaborate circles appear to have better mating success, which drives the evolution of this elaborate artistry.

While the little Japanese pufferfish might be a piscine Picasso, he is not alone among fishes in using sand as a medium for aesthetic expression. Like the Australian bowerbirds famous for the elaborate structures they build to attract and impress females, many cichlid fishes also build bowers to improve their mating prospects. The comparison to bird bowers is not superficial. For much like those of their distant feathered relatives, the fishes' bowers function mainly as way stations for display, courtship, and spawning. Almost as soon as eggs are laid, females pick them up in their mouths and move to a safer brooding location.

How does a fish build a bower? Lacking grasping appendages, male cichlids must use their mouths to pick up and deposit sand, and their fins to waft it. Each species constructs a bower of different design, from simple depressions, to arenas with radiating spokes, to volcano-shaped sand castles projecting a foot or more from the bottom with a courting platform at the top. The height or depth of these aquatic edifices advertises the male's health and the quality of his genes. The engine that drives these efforts by males is choosy females, who can detect subtle variations in a male's quality. When females preferentially mate with males with superior architectural skills, building prowess is favored over the generations.

Male sticklebacks also use their mouths to build their mating bowers (whose U shape is strikingly like that of some bowerbirds), but they have an extra tool to aid them. They produce a sticky, mucus-like substance in their kidneys. When it comes time to decorate house, a male extrudes this threadlike glue through his cloaca and uses it to bind together pieces of leaves, grass, and algae filaments to his nest. In a study of three-spined sticklebacks off the west coast of Sweden, Sara Östlund-Nilsson and Mikael Holmlund from the University of Oslo noted that males chose algae of deviant coloration to adorn the entrances to their nests—an apparent decoration. When the researchers placed bits of shiny tinfoil and bangles nearby, the male fishes didn't waste any time plucking them up and using them to decorate. Flashier nests attracted more females despite their being less camouflaged against predators. Humans and bowerbirds are not the only creatures who like a bit of bling.

Fake Orgasms and Sperm Drinking

Artistry is just one way to win a mate. Another is that familiar old tactic: deception. As we've seen, fishes are not above a little trickery.

In the case of a female brown trout, it takes the form of a fake orgasm. Having made a depression in the sand that serves as a nest, a female trout normally releases her eggs with vigorous body quivering in the presence of an amorous male. Caught up in the moment and seizing his opportunity, the nearby male follows the female's lead, furiously quivering while releasing his sperm into the water. But sometimes he has been duped. Her own quivers were eggless. It isn't clear why a female trout might practice this deception. One possibility is that she is testing the male's vigor. Or maybe she has decided that he is not the one for her and she seeks to draw in other males in hopes of finding a better father. This sort of reproductive conflict is common in nature. A male can afford to fertilize all of a female's eggs with his abundant, cheap sperm, and still have more left over for other females. A female, in contrast, may do better by having several sires for her precious eggs, improving her odds that some will be fertilized by the best-quality males.

Female cardinalfishes have a deception of their own. Or do they? Males protect the female's eggs by gingerly carrying them in their mouths. It's a major self-sacrifice on Dad's part, for he must forgo eating during this critical period of the reproductive cycle. Sometimes it is just too much for a starving male to bear, and he has been known to swallow the whole egg bunch in one gulp. In a ploy to lessen the chances of this unwelcome outcome, cardinalfish moms will lay a number of yolkless "dummy eggs" along with the real ones. The theory is that these fakes trick Dad into thinking he has more future offspring in his mouth, and that the clutch is therefore more worthy of careful protection. I find this interpretation unconvincing, for it assumes an exploitative partnership when a virtuous one might serve just as well. Perhaps we'll discover that females produce so-called "trophic eggs" to reward males for their investment, with males discriminating faux eggs from fertilized ones, eating the former and saving the rest. After all, the fertilized eggs contain his investment, too.

Among the diverse cichlids of Lake Malawi, it is the males who tweak their partners with eggy stimuli, this time in the form of egg mimicry. A female deposits her eggs on the substrate before collecting them in her mouth. As an aid to fertilization, the male wears a tattoo of yellow spots on his anal fin, giving the 3-D impression of a small group of eggs. The little cluster is irresistible to females. Drawn close to the male's reproductive organs, the female inhales most of the sperm he ejaculates, where it may better fertilize the eggs already in her mouth. This has been described as an apparent visual deception, but I doubt the male's "egg" spots act as a trick so much as a stimulus. Procreation is as vital to females as to males, so perhaps the female is not so much deluded as excited by the male's seductive visual signal.

Oral sex plays a more direct role in fertilization for the armored catfishes, *Corydoras,* a popular genus of the aquarium trade. The female drinks semen directly from the male by attaching her mouth around his genital opening. The sperm pass rapidly through the female's digestive tract, and she releases them onto a freshly extruded cluster of about thirty eggs that she holds between her pelvic fins.

I doubt I am the only one wondering how the sperm are not destroyed by the female's digestive enzymes. It helps that the semen is shunted through the female's gut at astonishing speed. A Japanese research team timed the sperm's passage in twenty-two females by squirting a puff of blue dye into the female's mouth at the moment she inhaled the male's sperm. Then they waited for a cloud of blue to emerge from the female's anus. (Surely, this loss of privacy is exacerbated by indignity.) They didn't have to wait long. Average time: 4.2 seconds!

These little catfishes have another adaptation that may facilitate rapid sperm passage and survival in the female's gut. They use intestinal air breathing, gulping air at the surface and passing it quickly through their intestines. So the digestive system of *Corydoras* appears preadapted for passing sperm quickly and unharmed through their bodies.

Why would fishes resort to such a dramatic means of fertilizing their eggs? For one thing, both parents know whose genes are being combined, which is good if you've chosen your mate carefully. A related advantage for the male is that he knows it is his sperm and only his sperm fertilizing the female's eggs. Whatever the ultimate benefits, sperm drinking certainly works for catfishes, for it is believed to occur in as many as twenty species.

A female's gut might not be the most bizarre place to combine sperm and eggs; what about the innards of an invertebrate? One of the most elegant symbioses in the oceans involves a curious sexual arrangement between bitterlings—small fishes of European streams—and mussels. When it comes time to mate, a female bitterling finds a suitably-sized mussel of the *Unio* genus in which to lay her eggs. How does she get her eggs inside a tight-lipped mussel? The bitterling mother-to-be uses a long, hoselike egg-laying tube to insert her eggs into the bivalve's siphon, itself a tubelike structure used by the mollusk for filtering water and food. Once the eggs are inside the mussel, the male bitterling releases his sperm near the siphon entrance, some of which is inhaled by the mussel. Over the ensuing days, the bitterling's fertilized eggs hatch and develop in the safety of their molluscan cloister.

That sounds all well and good for the bitterlings, but how, if at all, does the mussel benefit from being a bitterling's brood receptacle? The answer is that the mussel does not release the bitterling fry until its own eggs have properly ripened. The mussel's eggs adhere temporarily to the baby fishes, which provide a convenient egg dispersal service for the mollusk. Like the clasping seeds of certain plants that stick to animals' fur (and our clothes) for a free dispersal service, the mussel's eggs get a head start on finding fertile new terrain to establish themselves. One good deed deserves another.

Manipulation with a Feminist Twist

I've seen pictures of a female bitterling siphoning her eggs into the clam's inner sanctum like a fueling nozzle at a gas station, and I wonder if she is cognizant of how eccentric is her mode of reproduction. It seems she must know instinctively what to do, unless she learns it by watching other spawning females. The mating behavior of a male Atlantic molly seems less instinctual, for he varies his conduct according to the social setting; specifically, he deceives rival males by pretending he's attracted to someone else. Male mollies have an intromittent organ called a gonopodium—a fleshy appendage supported by a bone, which functions as a penis. A male can signal his sexual interest in a female by nipping her and thrusting his gonopodium toward her. In a study led by Martin Plath, male Atlantic mollies were presented with pairs of females with and without an audience. First, males were placed singly in tanks with a pair of females and allowed to show their preference. In the next phase of the experiment, males were again presented with the same two females, but half of the males were now being watched by a rival male placed in a transparent cylinder at the back of the tank.

Control males who lacked an audience (the cylinder at the back of the tank was empty) showed no change in female preference. However, almost all of the males who were being watched by a rival started to act as if they preferred the formerly non-preferred female. They made the switch from a larger to a smaller Atlantic molly, and they even made the switch from a female of their own species to a close relative, the Amazon molly.

It is thought that a male molly does this with the purpose of steering his rival's attentions away from a more desirable female.

Earlier studies found that male mollies are influenced by the preferences of rivals, including a switch from an Atlantic to an Amazon molly. The ploy may act to further reduce sperm competition, because nearby males may use this public information to copy the original male's feigned mate choice. By deflecting rival males' attentions to a different female, the first male improves his odds in the sperm lottery with his preferred mate by raising the proportion of her eggs he is likely to fertilize.

There is a feminist twist to this story of molly manipulation. Unlike their close cousins the Atlantic mollies, female Amazon mollies are an all-female species; there are no males. There are a few other all-female species among reptiles, amphibians, fishes, and birds. These species are referred to as parthenogenetic because no sperm is required to fertilize their eggs. But the situation is even more peculiar in Amazon mollies, because they can produce fertile eggs only if they mate with a male Molly of another species. Although the mating act is necessary to trigger pregnancy, it's a case of "sperm donor lite" for males, whose sperm do not actually fertilize the female's egg. Therefore, male mollies who mate with Amazon mollies are the victims of an immaculate deception.

You may be wondering why natural selection would tolerate males mating with females with whom their sperm reaches a dead end. These males appear to be benefiting by boosting their desirability to female Atlantic mollies. Several fishes, including mollies and their close relatives the guppies, are known to be trend conscious, and Atlantic females often copy the choices made by their Amazon counterparts.

Well-Hung Fish

Mollies are one, albeit bizarre, example of many fishes in which fertilization is internal. Most fishes mate without penetration, but there are many exceptions. All male elasmobranchs (the sharks and rays) have claspers, paired organs that the male inserts into the female's genital opening for sexual intercourse. Among bony fishes, males of the family that includes guppies, mollies, platys, and swordtails all possess a gonopodium.

Most of the time the gonopodium is directed backward, but when needed it can be swung in different directions. I remember an undergraduate animal behavior lab in which we recorded how often aroused male guppies made "gonopodial swings" and "sigmoidals"—an S-shaped body posture signaling readiness for sex. The flamboyantly colored males waved their gonopodia about like unruly wands, apparently to impress females. Although the guppies were tiny—most between one and two inches long—their gonopodia were about a fifth of their body length (I'll let you do the math), making them easily observed and accounted for by students and female guppies.

The priapium fishes of the appropriately named Phallostethidae (translation: chest-penis) also have penetrative sex. These are small (up to 1.4-inches), humble-looking creatures, numbering twenty-three species. Partially translucent, they live in brackish waters of Thailand and the Philippines. They are named for a muscular, bony copulatory organ, the priapium, found under the throat of males. Yes, you heard me right. In some species the priapium is even accompanied by a fully functional testicle. Another feature of the priapium is a serrated hook, the ctenactinium, which grasps and holds on to the female during the sex act. Careful anatomical study confirms that this remarkably complex apparatus is derived from the missing pelvic girdle and pelvic fins.

It is a testament to the importance of sex that evolution would find it within itself to ditch a pair of useful fins in exchange for a copulatory aid. It's also a nod to the mystery of life that ancestors of these fishes seemed to get along perfectly well without a priapium. Nobody knows why these penises have migrated toward the fishes' head end. Dare I guess that having his penis located near his eyes makes a male priapium fish more likely to achieve an accurate approach for insertion?

What do female fishes think of male fish parts? More to the point, does size matter in the fish world? It seems to for mosquitofishes, in which the gonopodium can extend to 70 percent of the length of a male's body. The biologist Brian Langerhans from Washington University in St. Louis tested the "size matters" theory by placing a female mosquitofish into a tank and projecting an image of a male on each side. One male's gonopodium had been digitally manipulated to appear longer than the other. In every trial, the female swam toward the male with the longer organ. But ever-efficient nature poses restrictions on extravagance, and being well-hung comes with at least one liability for a male mosquitofish. Just as a peacock burdened with a tail two feet longer than the competition would be more likely to fall victim to a predator before he even had the chance to breed, mosquitofishes with larger organs are more vulnerable to their enemies. Big gonopodia produce more drag in the water, making their bearers easier to catch. Fittingly, males living in predator-infested lakes have smaller gonopodia than males in safer waters.

By focusing on fishes with penetrative sex, I do not wish to demean those numerous fish species whose eggs and sperm are released into the water for so-called external fertilization. That style of breeding manifests in countless ways among fishes. Let me briefly provide one example: the complex nesting and mating behavior of sea lampreys, which defies the "primitive" stereotype these ancient jawless fishes are labeled with. Like salmons, they are anadromous—having both marine and freshwater stages in their life history. At spawning time they surge upstream to build an oval nest two to three feet in diameter. A mated pair use their suction mouths to lift or drag stones to a pile upstream from the nest. In mating, the female grasps a rock with her mouth, the male grasps the female behind her head, then twines his body around hers, and then they both vibrate vigorously. This motion stirs up fine sand that sticks to the releasing eggs, helping them sink into the nest. Next, the parents separate and begin removing stones from above the nest and placing them on the downstream side, which performs two functions: loosening sand that further covers the eggs, and shoring up the nest cavity to secure the eggs in place. The parents repeat the whole process until all eggs are extruded. This odyssey has a Romeo and Juliet ending: the pair are so exhausted by the end that they soon die.

As usual, what we know of the sexual behavior of fishes is only a fraction of what goes on. Of those species that have been studied, many were in artificial environments, which has the advantage of convenience, but which may have the unfortunate effect of suppressing sexual behaviors readily expressed in the wild. Captive lemonpeel angelfishes, for example, do not show courtship activities normally associated with harem maintenance. We may wonder what marvels await discovery, or shall remain forever hidden in the depths.

One thing we do know is that, for many fishes, reproduction does not end with sex. There are young to raise, and that leads to some creative problem solving.

Parenting Styles

No one is useless in this world who lightens the burden of it for anyone else.
—Charles Dickens

When I was eight my teacher showed us a film about salmons who make the epic journey from the ocean back to their natal stream to spawn and die. We had to write a report on the film. My mother kept mine. Here's an excerpt:

> There had to be many eggs because salmon have many enemies and all the eggs would be eaten up. After a few weaks or so there were about 15 eggs left. By a month [the hatchlings] had fed a lot and grown big enough to recognize that they were salmon. Suddenly a big object came swimming towards them all the little fish swam for their lives. But most of them got caught and eaten up it was a big pike.

To my best recollection, that film instilled the impression that a salmon's life is an unmitigated, earnest struggle, although I did have the good sense to conclude that while "mating you would think they are fighting but actually they are thoughrally injoing it." Notwithstanding a young boy's sometimes laughable attempts to express facts, that film conveyed at least one other misconception about the lives of fishes. Despite what we're taught—that salmons complete their life cycles and die after spawning in their home streams—in fact, some males and many females turn right back around and return to the sea to regain their normal body condition and resume adult life; years may pass before they respond again to the reproductive urge.

The film also led me to think that fishes do not look after their babies. In fact, caregiving has evolved at least twenty-two different times in fishes. About one in four of all fish species—some 8,000 species—devote at least some form of caregiving to their offspring. The effort ranges from protecting eggs to looking after the young through their most vulnerable first few weeks of life. Many fishes, including sharks, are viviparous, that is, giving birth to live young. Some sharks have a placenta to nourish developing embryos through an umbilical cord before they are live-born.

Despite these reminders of mammalian reproduction, fishes do not feed their babies milk as mammals do. Nonetheless, some species produce body substances that act as food for their young. The best known is the discus, a popular aquarium cichlid from South America. During several weeks of caring for their developing fry, discus parents allow their babies to feed on the protective mucus layer that covers their bodies. This isn't just any old mucus: it is grown from specialized thickened scales on the parents' flanks. It's a personal nutrition service with an immunity booster: the mucus is enriched with antimicrobial substances that help protect the babies from infections. Scientists are finding that immune boosters are not rare among fishes. A new family of peptide antibiotics called *piscidins* (translation: fish-related chemical compounds) was isolated from fish mucus.

Scrumptious slime is not the only milk substitute in fishdom. Recall those unfertilized "trophic eggs" that female cardinalfishes produce for the mouthbrooding males? Many sharks provide trophic eggs as an added food source to their developing embryos prior to birth. One species of catfish in Lake Malawi is known to feed trophic eggs to its free-swimming juveniles. The youngsters position themselves near the mother's vent and eat them as she extrudes them into the water. Caviar on the fly.

Egg Protection

Before baby fishes are born, expectant parents take on the role of protecting the eggs. One approach is to guard the eggs by chasing off intruders. True to their pugnacious nature, damselfishes are vigorously protective parents. On an hour-long snorkel at a small reef off of Key Largo, Florida, I saw only a handful of aggressive interactions between fishes, and almost all involved chases by yellowtail damselfishes. Tierney Thys, a world expert on fishes, describes an encounter with a damselfish protecting an egg clutch. The five-inch fish grunted repeated warnings at Thys as she approached for a closer look. Failing to repel the giant diver, the damselfish darted in, "snatched up a large strand of my hair in its tiny teeth, and yanked backward . . . so hard, that I involuntarily yelped in pain, a cry that was immediately followed by spluttering as I flooded my mask laughing."

Alternatively, parents may hide their eggs by constructing various types of nests or shelters, including cavities, elaborate structures made from plant materials, and rafts of bubbles blown from specialized saliva. Whitetail major damselfishes take a white-glove approach. The mated pair clean their egg-laying site by sandblasting it. The parents pick up sand in their mouths and spit it with force against the chosen rock face. Then they fan the area with their fins. Finally, they remove any sand grains that remain stuck to the rock face by plucking them off with their mouths.

A more fins-on approach to improving egg survival is to carry one's eggs, which may take the form of transporting them in the mouth, or carrying them in a pouch—for which male seahorses are renowned. In the amazingly camouflaged ghost pipefishes, native to the tropical Indian Ocean, the female's pelvic fins fuse together, and the resulting pouch functions as a cradle; in "true" pipefishes, relatives of seahorses, it is the male who has the pouch. The male humphead of New Guinea carries his mate's eggs dangling like a bunch of grapes from a projection on his forehead. One bottom-dwelling catfish from Guiana actually wears its eggs. The parent rolls in the egg mass so that the eggs adhere to the skin, where they are overgrown by a new layer of skin until the embryos are developed enough to emerge from their unconventional wombs.

South American cichlids called banded acaras lay their eggs on a loose leaf they have carefully chosen. The male and female of a mating pair often "test" leaves before spawning: they pull and lift and turn the candidate foliage, trying to select leaves that are easy to move. After spawning, both parents guard the eggs. When disturbed, the parent acaras often seize one end of the egg-carrying leaf in their mouth and drag it hastily to deeper and safer locations.

I have special admiration for the spraying characin, named for their eccentric form of egg-caring behavior. Rather than lay their eggs on underwater leaves as the banded acaras do, these athletic fishes deposit them in the air, on overhanging leaves. The parents-to-be line up vertically just below the waterline; then, on some subtle, split-second cue, they leap upward, usually in unison, to the chosen leaf. Each leap culminates with the two fishes turning upside down and depositing sperm and about a dozen eggs. Talk about good timing! In this manner, several dozen translucent (and well-camouflaged) eggs end up adhering in a cluster on the target leaf. I've read that leaps can be four inches high, but watching a film of the behavior indicates that characins can jump much higher. They can also buy more time to deposit their goods by clinging to the leaf for several seconds.

The incubation period is very short, which is just as well

because Dad must work in overdrive to keep the eggs moist. He does this by firing water onto the egg masses with a skillful tail-flick. It must be an exhausting job, for splashing is performed at one-minute intervals during the two to three days until the eggs hatch and the newborns drop into the water.

When I encounter curious animal behaviors like this I can't help wondering how they arise. How does a fish go from laying and caring for eggs in the water to the outlandish custom of depositing them on a leaf and spraying them with water? Surely, the answer must be: gradually, and in stages. In some ancient characin environment, perhaps there was a visual predator who was thwarted when the characins laid their eggs on underwater leaves. Then, other predation pressures might have compelled an enterprising pair of characins to reach up with their loins and deposit their by now sticky eggs on low-hanging leaves just above the surface. In time, perhaps driven to greater heights by more determined aquatic predators, the characins developed their leaping skills. At every step of the way, some advantage must have accrued, else the behavior would not have been genetically favored in the population.

Spraying characins are not the only fishes that place their eggs out of water. Various intertidal species, inhabiting the zone between high and low tide, make a specialty of airborne egg brooding. Pricklebacks, gunnels, and wolf eels coil their elongate bodies around their egg cluster as the tide recedes, trapping a small pool of water in which the eggs sit. It says something about the virtue of parental dedication that a fish will lie for many hours, exposed to air, to protect his or her future offspring.

Further strategies for protecting eggs above the waterline include covering them with seaweed, burying them in the sand, and hiding them among rocks. There must be advantages: higher incubation temperatures, higher oxygen concentrations, and lower predation.

Gargle but Don't Swallow

The most ingenious method fishes have devised for protecting their young at their smallest and most vulnerable is to carry them in their capacious mouths. Mouthbrooding, which includes carrying eggs or free-swimming babies, occurs in at least nine fish families on four continents. In the latter case, when a family is threatened the parent fish may signal danger by backing up slowly with the head down. The young approach and are sucked into the parent's mouth, only to be let out again when the danger has passed. It looks like vomiting in reverse.

Cichlid fishes are mouthbrooding specialists, with 70 percent of the 2,000 known species using their mouths for day care. The great diversity and success of the cichlid family may be due in part to this adaptation. It is probably because one can only fit so many babies inside one's mouth that brood sizes of cichlids are smaller than those of many other fishes. But smaller families are well compensated by a higher proportion of fry surviving infancy.

Among the best-known groups of mouthbrooders are members of the genus *Betta,* which numbers more than seventy species. Some bettas protect their young in bubble nests, which might be an evolutionary precursor to mouthbrooding. Bubble nests work well in stagnant water where bubble-nesting bettas live. They keep the eggs and developing fry together, safe, moist, and close to the oxygen-rich atmosphere. But in moving water such as a stream, a bubble nest is very difficult to maintain. Parents manipulate eggs with their mouths during the construction of bubble nests, so it is just a short evolutionary hop to holding the eggs in the mouth. One can imagine an ancestral fish, exiled into a new stream habitat, watching his bubbles float away, discovering through desperation that things go better if he puts his money where his mouth is.

Mouthbrooding has other advantages. A bubble nester is tied to the nest and cannot venture far from the homestead without risk of losing eggs or fry. Mouthbrooders can move at will to keep themselves and their broods safe. And the eggs are kept well oxygenated by moving a current of water over them with every breath.

Mouthbrooding isn't just clever, it is virtuous. Typically, mouthbrooding parents stop eating during the entire period that eggs or young are being harbored in the mouth. That's no small thing, for the mouthbrooding period may last a month or more. No wonder that mouthbrooders have been known to starve to death.

It gets nobler still. The parent continues to take food into his or her mouth, but it isn't swallowed—at least, not by the parent. These morsels are instead fed to the young while they are holed up in the doting parent's mouth. For instance, as revealed in a study of wild blunthead cichlids in Lake Tanganyika, mothers swim off to a quiet area of the lake to mouthbrood for about thirty-three days. They take no food into the gut during this period, but their browsing rate increases to meet the needs of their growing offspring. For sheer restraint, that has to rank among the highest in the animal kingdom.

Good Dads

Blunthead cichlids notwithstanding, guess who does the lion's share of the child-care work among fishes? Dad. In contrast to the pattern on land, where mothers often shoulder most of the parental duties, the roles are usually reversed in fishes. Unavoidably, females continue to bear the costs of egg production, but it is usually males who take over from there. So it is the father betta who builds the bubble nest and who guards the developing eggs until the fry hatch. Sensing danger, it is he who shakes his pectoral fins close to the surface of the water, creating a wave that can be detected by young, who then swim to him for the safety of his oral cavity.

So predominant is the male's role in mouthbrooding that in some cases his face has evolved to better suit the task. When the heads of nine cardinalfish species were closely examined, males were found to have longer snouts and jaws than females. The investigators further reasoned that the mouth's role in brooding young constrains another important function of these fishes' mouths: respiration. With precious space taken up by dozens of offspring (all of whom are themselves drawing oxygen from the water), the caregiver's oxygen intake is compromised. This has spawned predictions that cardinalfishes may expect to face a grim future. Says David Bellwood of the School of Marine and Tropical Biology at James Cook University in Queensland, Australia: "Mouthbrooding makes them more vulnerable to the effects of climate change. As ocean temperatures warm, these fish will need to breathe more—and the last thing they need is having a mouthful of offspring when they need oxygen."

The champion fathers among fishes are the seahorses and their close relatives, the pipefishes. These males come about as close to being pregnant as it gets. The female releases her eggs into the male's abdominal pouch, where he fertilizes them and carries them until they hatch. "Birthing" involves contractions and contortions to expel the young from the pouch.

There are significant pluses to this "pregnant dad" system. From a purely reproductive standpoint, dad gets the double benefit of (a) paternity assurance, and (b) higher numbers of his children surviving to independence than if they were simply cast off to face the hazards of sea life alone. Paternity assurance is no trifling matter in nature. Whereas mothers—in return for the substantial energetic costs of pregnancy and (where applicable) child-rearing—know with certainty that their young are their own, fathers can rarely be sure. Ironically, this male-centered guardianship system actually shifts parental uncertainty to the female. Genetic analyses reveal that rates of monogamy in male seahorses can be as low as 10 percent, and males have been found carrying eggs from as many as six females. There is, however, evidence that females are also playing a numbers game by contributing eggs to more than one male pouch.

Helpers

Paternity uncertainty is only one obstacle to realizing your reproductive potential. A shortage of resources to settle and grow a family is another problem. Nest sites, food availability, and suitable mates may all be in short supply, and this can lead to major compromises.

As a graduate student, I met once a week with a small group of behavioral ecologists to discuss the latest research on cooperative breeding in birds. Such is the diversity of this phenomenon that entire courses, and several books, have been devoted to it. Cooperative breeding happens when one or more nonbreeding adults forgo the opportunity to breed in favor of assisting with the child-rearing duties of another adult pair of birds. The breeding pair are often, though not always, the helpers' parents. Cooperative breeding is known from several hundred bird species, including babblers, jays, kingfishers, and hornbills.

I took that course in 1989. Curiously, nobody ever mentioned cooperative breeding in fishes, even though it had already been documented years earlier in daffodil cichlids (more on them soon). Cooperative breeding is currently known from far fewer fishes (only a dozen or so species) than either birds (about 300 species) or mammals (120), but the relative secrecy of fishes' lives means that many others may remain undiscovered.

The best known cooperative breeding fishes are those innovators, the cichlids. Helpers perform a variety of tasks relating to the care and protection of eggs and young, such as cleaning and fanning eggs and fry; removing sand and snails from the breeding area; and defending the parents' territory.

Helping behavior in birds and mammals is thought to have evolved via kin selection. If opportunities to raise a family of one's own are limited, say, due to a lack of suitable nesting sites, then it makes more sense to help relatives than to just bide one's time and do nothing. Helping increases the helper's genetic fitness by benefiting kin who share the helper's genes. Helping also provides valuable training. As a future breeder you are more likely to succeed in the lively arts of nest building, incubation, feeding of young, and nest defense if you have first completed a full apprenticeship.

That said, you should still strike out and raise your own young if circumstances permit. For birds, supportive evidence comes from a study of Seychelles warblers, who only began to show helping behavior after all high-quality nesting sites were occupied following introduction to a new island. Once the landgrab was complete, compromise followed.

Do fish helpers also help because they don't have any better options? Swiss researchers from the University of Bern set out to test this so-called ecological constraints hypothesis in an elaborate captive study of daffodil cichlids taken from the southern end of Lake Tanganyika. Daffodil cichlids are the darlings of researchers studying cooperative breeding in fishes. They are small (the largest just under three inches), graceful fishes with large eyes, pinkish-yellow bodies, and long, wispy fins bordered with sky blue. Their social lives are no less colorful. Nest helping includes digging to remove sand from breeding shelters; defending the nest by such maneuvers as mouth fighting, biting, ramming, spreading of fins or gill covers, head-down displays, and S-shaped body bends; and mollifying higher-ranking fishes with various submissive behaviors, including tail-quivering, "hook displays," and fleeing.

In the Swiss lab, thirty-two pairs were set up in thirty-two breeding compartments of a 1,900-gallon circular tank, with a "dispersal compartment" adjoining every four breeding compartments. In addition to ample amounts of sand, each breeding compartment and half of the dispersal compartments contained two flowerpot halves serving as breeding shelters. Each breeding pair (sixty-four fishes in all) was also assigned a mixed pair of helpers, one larger than the other, both smaller than the breeders: that's sixty-four helpers in all. Helpers were trained to swim through small slots in the Plexiglas partitions dividing the breeding from the dispersal compartment. These slots were too small for the larger breeders to swim through.

Despite the disorientation of intercontinental transport, the fishes soon adjusted to their new surroundings. One breeding pair produced a clutch of eggs within five days of arrival, and all but one of the thirty-two breeding pairs produced at least one clutch of eggs during the four-and-a-half-month experiment.

Did helpers help or start their own families when they had the opportunity to do so? They started families. As predicted by the ecological constraints hypothesis, helpers who had access to a breeding shelter defected to the available dispersal compartment, paired up with another helper, and produced their own young. Larger helpers provided less help to their assigned breeding pair, and large helpers with access to their own breeding shelter grew more in size than those without a breeding shelter, which suggests that these fishes can strategically control their body size according to their breeding status.

Of the helpers who bred, none mated with the other helper originally assigned to their breeding compartment, probably because the second helpers were smaller and thus perceived as less suitable for mating than one of the larger helpers from an adjoining breeding compartment. No breeding occurred in the dispersal compartments without breeding shelters, which illustrates the importance of proper breeding supplies.

This cleverly designed study shows that for daffodil cichlids, as for many birds, helping is a compromise due to limited resources in the environment. It reminds me of volunteering at an organization or interning at a business as a prelude to being taken on as an employee or to starting one's own company.

Helping others raise young is virtuous, but some daffodil cichlid helpers may be less virtuous than others, getting more out of the deal than an apprenticeship and an indirect genetic investment. A genetic analysis of wild daffodils at Kasakalawe Point, Zambia, found that while breeding females were the mothers of virtually all of their offspring, breeding males were siring less than 90 percent. In over a quarter of clutches, male helpers were getting in on the action. Genetic data collected from groups of daffodil cichlids from Lake Tanganyika revealed mixed parentage in four of five of the examined groups.

This is not all bad news for the dominant breeding male—who in any event usually remains oblivious to the helper's indiscretion. Knowing they have a higher genetic stake in the clutch, transgressing helper males show more vigorous defense against egg predators than do subordinates not participating in reproduction, and they tend to stay closer to the breeding shelter. When helpers are temporarily prevented from helping, other group members compensate by increasing defense of the territory. After being returned to the nest site, the thwarted helpers increase their helping behavior, even though the scientists find no evidence of their being punished as slackers by the breeding pair.

These dynamics are not alien to our human societies. Despite societal norms about monogamy and sexual fidelity, things often get messy; else we wouldn't have terms like "cheating," "cuckold," and "paternity testing." Or fostering and adoption.

Freeloaders

The virtuous act of nest helping has paved the way to a form of transgression in the world of fish parenting. It is what biologists call brood parasitism.

As with helpers at the nest, brood parasitism is best known among birds. It is the art of laying one's own eggs in another's nest. Practiced also by certain fishes, amphibians, and insects, brood parasitism is an evolutionary freeloading strategy in which someone else does the work of protecting and raising one's young. Many avian brood parasites will remove a host egg when they lay one of their own in a nest. In cases where the host nestlings are significantly smaller than the parasite nestling, the parasite gets most of the food, and the host's own chicks may starve to death. In its grimmest form, some brood parasites, particularly cuckoos, will eliminate eggs or newly hatched nestmates, either by ejecting them from the nest or killing them with a sharp hook on the baby cuckoo's beak, which falls off after a few days. Others, such as the giant cowbird, are not known to harm the oropendola or cacique chicks whose nests they parasitize, and there is evidence for a beneficial trade-off when the interloper chicks pluck parasitic botfly maggots from their nestmates.

Among fishes, the best known examples of brood parasitism occur in the large African lakes where fish social behavior has found some of its most elaborate expression. In Lake Malawi, a research team from Penn State University found evidence of brood parasitism in eleven of fourteen kampango catfish nests by one of the lake's most common catfish species, an endemic named the *bombe* by locals. Parasitized kampango nests held bombe young almost exclusively, and these were protected by kampango adults until they were about four inches long. In the kampangos, both females and males feed their young. The mother provides trophic eggs, for which the young fishes gather expectantly around her vent. The father collects invertebrates from the surrounding habitat, brings them back to the nest, then distributes them to the hungry offspring through his gill covers. In the parasitized broods, bombe babies fed side by side with kampango babies. As yet, nobody knows whether the bombe young instinctively know the feeding regime of their adoptive parents, or whether they learn it.

Parasitism of kampango catfishes by bombes might be more the exception than the norm. Before making the brood-parasitism observations in early 2007, Jay Stauffer had logged more than 1,600 hours of diving in Lake Malawi, but had never witnessed this behavior. And it isn't as if bombes are habitual freeloaders of kampangos; they will also care for and rigorously defend their own broods. Stauffer was bitten on the hand when he trespassed too close to videotape a bombe nesting area.

At least the bombes are somewhat civil in their parasitic relations with their kampango hosts. Five hundred miles northwest, in Lake Tanganyika, aptly named cuckoo catfishes spawn right above breeding cichlids—and the cichlids dutifully mouthbrood the catfishes' eggs and young. Adding insult to audacity, the catfish's eggs hatch earlier than the host's eggs, and once their yolks have been absorbed, the baby catfishes start eating their cichlid broodmates. When the zoologist Tetsu Sato of Kyoto University reported this in 1986, it was the first known fish example of true brood parasitism—in which the young of the imposter species are totally dependent on another species' parents.

If there is one overarching conclusion we can draw from the current science on fishes, it is this: fishes are not merely alive—they have lives. They are not just things, but beings. A fish is an individual with a personality and relationships. He or she can plan and learn, perceive and innovate, soothe and scheme, experience moments of pleasure, fear, playfulness, pain, and—I suspect—joy. A fish feels and knows. How does that knowledge mesh with our relationship to fishes?

PART VII

FISH OUT OF WATER

I, a many-fingered horror of daylight to him,
have made him die.

—from "Fish," by D. H. Lawrence

It isn't easy being a fish, especially in an age of humans. Humans have fished since time immemorial. Eons before livestock were corralled in fences, fishes were being caught on hooks and in nets. The oldest fishhook ever found so far dates to between 16,000 and 23,000 years ago. The earliest known fishing net was discovered in 1913 by a Finnish farmer while digging a ditch in a swampy meadow; made of willow fibers, the almost 100-foot-long by 5-foot-wide net was carbon dated to 8300 B.C.

I doubt the early fishermen wielding their hooks or tossing their nets in the shallows fretted about catching all the fish in what to them must have seemed a boundless ocean stretching beyond the horizon. And they needn't have. Indigenous fishing communities have been living in harmony with wild fishes for as long as recorded time. Long-term survival demands a sustainable balance between their needs and the fishes'. It's a different story in the modern world, where fishing is not just done for subsistence but for profit.

Well into the twentieth century, it was widely believed that the world's waters contained an unlimited "supply" of fish. A few years back I rescued an old book from an alleyway trash heap. In *Animal Life of the World*, published the year of my mother's birth,

1934, H. J. Shepstone writes: "Though every year fish are taken from the sea in millions of tons, yet there are no signs that this store will ever become depleted."

Much the same was said of the passenger pigeon. And we know how that turned out.

Mr. Shepstone failed to account for two trends that were already well in evidence in his day. The first is the steady growth in the number of humans on Earth. All else remaining the same, that growth translates into more consumption. Even if per capita consumption of fishes remained unchanged, about three times more fish would be eaten today simply because world human population has tripled since Shepstone's essay was published.

Today, consumption of fish has grown dramatically in the world's two most populous nations. The average Chinese citizen eats five times more fish than in 1961, and the average Indian more than two times. During that half century the populations of these nations have also more than doubled. According to the Food and Agriculture Organization, the average human in 2009 was consuming 40.6 pounds of fish, nearly double what she was consuming in the 1960s. In the United States, per capita fish consumption has remained fairly constant, which still represents a large increase because there are more Americans, and we are also feeding more fish to other animals we eat.

Any notion that these increases might reflect rising fish populations is an illusion. The opposite is true. Global fish numbers are shrinking, and the number of collapsed fisheries has grown steadily since 1950.*

Isn't that a paradox? How can humans be eating ever more fish when the fishes' populations are falling? "Anybody who thinks there can be limitless growth in a static, limited environment [like the oceans], is either mad or an economist," quips the British biol-

*This state of affairs is exacerbated by government subsidies of commercial fishing, to the tune of $35 billion per year worldwide.

ogist and television presenter Sir David Attenborough. This brings us to the second trend absent from Mr. Shepstone's calculus: the steady advance of technology. It has transformed commercial fishing. Today's vessels can track down fish schools with sonar, satellite navigation (or GPS), depth sensors, and detailed maps of the ocean floor. Some deploy spotter planes, others use helicopters. Nets of durable, lightweight synthetic fibers, some miles long, are cast into the sea. Purse seines a mile in length and 250 yards deep encircle schools of sardines, herrings, and tunas near the surface. The net is then drawn together at the bottom (creating a purse) for hauling on board. In longline fishing, lines with 2,500 or more hand-baited hooks, some stretching to over sixty miles, hang at various depths below the surface, or may be weighted to lie at the bottom, half a mile down. Huge winches haul the catch on deck.

The most destructive and indiscriminate of all fishing methods is bottom trawling. A trawler is like a lawn mower with a big weighted net to catch the cuttings. Armed with heavy metal rollers, these nets are dragged across the seafloor at depths of half a mile to a mile, indiscriminately scooping up everything in their path. A hundred years of structure on the bottom—corals, sponges, sea fans, etc., which provide vital spawning habitat for fishes—is seriously damaged or destroyed by one pass of a trawl net. Fishes of all ages and sizes, plus seaweeds, anemones, sea stars, and crabs, are removed or destroyed. The celebrated American oceanographer and TED Prize winner Sylvia Earle likens trawling to "using a bulldozer to catch hummingbirds."

The fishing vessels themselves are not so much boats as seaborne factories, complete with refrigeration and canning operations for storing their catch. If their cargo reaches capacity they can transfer the catch to collection vessels, avoiding a time-wasting return to port. They stay at sea for weeks or even months at a time. And there are lots of these factory ships—more than 23,000 weighing 100 tons or more—plying the world's oceans.

Commercial fishing in the modern era is like bobbing for apples with hands instead of the mouth. The fishes don't stand a chance. Today, how many we take is no longer limited by how many we can catch, but by how many there are left to be taken.

Reared

The alternative to catching wild fishes in the sea is to rear them in captivity. Fish-farming (a subset of aquaculture, which includes such practices as rearing crocodiles for their skins, cultivating mussels for pearls, and growing seaweed) is the fastest-growing animal-food-producing sector in the world, having gone from 5 percent of global fish production in 1970 to about a two-fifths of the total today.* Aquaculture works on the same principle as factory farming of land animals. Fishes are kept in highly dense conditions, fed on a rich diet formulated to maximize growth, then slaughtered and processed for human consumption. Instead of crates and battery cages, farmed fishes are confined to marine and freshwater net pens, or land-based tanks or ponds. On trout farms, densities can be as high as twenty-seven foot-long fishes per bathtub volume of water.

At first glance, aquaculture might appear to be a savior for fishes in the wild. The reality is more complicated. Paradoxically, the production of factory-farmed fish does not relieve the pressure on wild fish populations. This is because the primary food fed to farmed fishes is, well, fish. Human diners have a taste for carnivorous fishes, whose natural diet is smaller fishes. Most of the "prey fish" captured on the seas (think anchovies or herrings) are fed not to humans but to farmed fishes and to pigs and chickens on factory farms. More than half the world's fish oil produc-

*Aquaculture equals commercial fisheries in total seafood production today, but because fish production accounts for only about half of all aquaculture (seaweed alone represents over a quarter of aquaculture production), fish production from aquaculture is about 40 percent that from fisheries.

tion is fed to farmed salmons, and 87 percent is used in aquaculture. How many fishes does it take to grow other fishes to marketable size? It varies. According to one analysis from 2000, it takes between two and five pounds of "feed fish" to make one pound of farmed carnivorous fish such as salmon, sea bass, or bluefin tuna. Given the smaller size of feed fishes, many individuals must be used to maintain the farmed species.

The most notable of the uncelebrated feed fishes is a species you have likely never seen or heard of, and almost certainly haven't eaten. The menhaden (which actually defines four commercially targeted species) is a humble-looking fish inhabiting both the Atlantic and Pacific Oceans. About a foot in length, and with a classical oval shape, forked tail, and bright silvery scales, these filter feeders would be a suitable exemplar for "fish" in an illustrated dictionary. So many menhadens are caught by humans that the cultural historian H. Bruce Franklin dubbed them *The Most Important Fish in the Sea* in his aptly titled book. A catch cap on Atlantic menhadens imposed in December 2012 by the Atlantic States Marine Fisheries Commission reduced the 2013 haul by 25 percent, or 300 million. That represents a prior annual regional catch of 1,200,000,000 individual menhadens.

Like a third of the world's fish catch, menhadens are not eaten by humans. *Menhaden* derives from a Native American word for fertilizer. Their commercial use is to be reduced into oil, solids, and meal. Dead, dried, and then pressed, menhaden oil is used in cosmetics, linoleum, health food supplements, lubricants, margarine, soap, insecticides, and paints. Most of the menhaden meal—a product of pulverizing the dried carcasses of the fishes—is fed to factory-farmed poultry and pigs; some also goes into pet food, and feed for farmed fishes. One company, Omega Protein, was as of 2010 operating sixty-one ships, thirty-two spotter planes, and five production facilities, all dedicated to the conversion of menhadens into money.

While wild fishes are being fed to farmed fishes, the farmed fishes are already on someone else's menu: sea lice. *Sea lice* is a generic term for many species of parasitic copepods that latch onto the bodies of fishes and other marine creatures and feed off their living tissues. In the wild, sea lice do not pose a major threat. But in the artificial conditions of intensive confinement, where the next host fish is just inches away, sea lice thrive. As they chew through the mucus, flesh, and eyes of fishes helpless to escape them, sea louse heaven becomes farmed fish hell. Overall death rates of 10 percent to 30 percent are considered acceptable in fish-farming.

The nets that hold fishes in their sea pens do not prevent these rampant parasites from escaping. A female sea louse lays about 22,000 eggs during her seven-month life span, and they spread in clouds over miles of surrounding waters, wreaking havoc on wild fishes who find themselves in the vicinity of the farms. The lice are credited with causing massive die-offs of 80 percent of wild pink salmons on Canada's Pacific coast. Trickle-down effects impact salmon-dependent wildlife: bears, eagles, and orcas.

Crowded fish farm conditions give rise to other problems. These include viral and bacterial diseases, such as infectious pancreatic necrosis (IPN), viral hemorrhagic septicemia (VHS), and epizootic hematopoietic necrosis (EHN); toxic chemicals used to treat them; and concentrated fish wastes. All of these contaminate surrounding waters, affecting native fishes and their habitats. A single farm in Lake Nicaragua that raises tilapia—the most popular farmed fish flesh in the United States—equals the impact of 3.7 million chickens defecating in the water. Many farmed fishes escape through nets damaged by seals or storms, diluting the genetic viability of wild populations.

Not only are they less viable, captive-reared fishes are less resourceful than their wild cousins. Brains, like muscles, need to be used to develop normally. Free-living fishes must learn to find prey, and to recognize and handle it. But a mundane, unstimulating life of captivity stunts brain development and function. When hatchery-reared fishes are recaptured after release into the wild, their stomachs are often empty or filled with inanimate objects such as floating debris or stones that look like the pellets they were reared on. No wonder: the young fishes have had no chance to learn how to make a living in the wild. There is potential to address this with judicious captive training regimens. Aware of fishes' observational learning abilities, the fish behaviorists Culum Brown and Kevin Laland have used a video of another fish eating live food to teach naive, hatchery-reared salmons to forage on live, novel prey items. But it seems doubtful whether training large numbers of densely crowded captive fishes is economically or logistically feasible.

A Visit to a Research Facility

To gain some firsthand exposure to fish-farming, I visited the Freshwater Institute (FI), a small aquaculture research facility nestled in woodlands of the Potomac watershed near Shepherdstown, West Virginia. My host was Chris Good, a tall, pleasant man in his mid-thirties. He was hired at the Freshwater Institute after completing veterinary and doctoral degrees at the Ontario Veterinary College, University of Guelph, Canada, where he focused on the epidemiology of fishes.

FI's objective is to advance the sustainability of aquaculture through numerous avenues, including research to improve the welfare of farmed fishes. It operates on a smaller scale than a typical commercial fish farm. Chris showed me into the main warehouse containing about a dozen cylindrical tanks reminiscent of brewery vats. The din of machinery and pumps was so loud that we had to shout to hear each other. The largest tank—30 feet across and 8.5 feet deep—contained about 4,000 to 5,000 foot-long salmon smolts, each about fourteen months old. A porthole revealed layers of greenish-brown fishes gliding effortlessly in an eternal circle. Patches of silvery scales glinted in the dim light.

Automated feeders dispensed feed pellets into the tank every one or two hours depending on a predetermined feeding regime. Bags of fish feed were stacked against the warehouse wall. I looked at the long ingredient list, which included poultry oil, fish oil, vegetable oil, and wheat gluten. It did not mention any fish species' names, but menhaden was almost certainly in there. Chris opened a bag so I could see the small, deep-burgundy pellets, each about a fifth of an inch across, which reminded me of dry cat food. I tasted one. The consistency resembled a hard graham cracker. The taste was faintly oily and salty, but otherwise bland.

We visited small vats with hundreds of young salmon parr just an inch or two long. We discussed jaw deformities, diarrhea outbreaks, research protocols, and dominance hierarchies (among the fishes, not the employees). Our tour finished at the end of a building, where the fishes are slaughtered. At FI, slaughter is preceded by seven days of no feeding, which is intended to purge fishes of "off-flavor," which can accumulate in muscle tissue of fishes from certain rearing systems, and lowers palatability to consumers. Chris told me that some broodfishes used for egg production are starved for seven or eight *months* in the belief that this increases the quality of their eggs, which he considers abhorrent from a welfare perspective. Chris showed me the holding tank to which fishes are transferred just before they meet their end. It's a stainless steel affair about eight feet long, rectangular at the deep end, with a midsection that narrows into a funnel at the business end. Mounted onto the funnel is a pneumatic device that delivers a stunning blow to the fish's head when it is forced to swim through the funnel; simultaneously, a sharp blade flicks out on each side to slice open the gills for bleeding out. Chris reports that the device is highly effective; on the occasions when a fish is not killed, such as when it enters the funnel facing the wrong way, or when it is upside down, a worker at the spill trough beyond the stunner-killer uses a handheld club to whack the fish on the head. He cautioned, however, that his facility's slow pace of killing helps keep the slaughter operation running smoothly, and that the story could be different in larger industry settings.

Dying to Be Eaten

Commercial fish stunners are the state of the art in fish killing. Most of the great masses of fishes killed for our consumption die differently. Out at sea, a single purse seine haul may contain half a million fishes if they are herrings, or if they are a larger species, like Chilean jack mackerels, then a net can contain a hundred thousand. Fishes caught this way are crushed by the weight of thousands of others as the net is tightened and hauled to the surface to be winched on board. Sometimes a submersible pump is lowered into the purse to suck the fishes up like a vacuum cleaner, then deposit them into dewatering boxes, then into holds below deck. Any fish who survives these events will most likely die from oxygen starvation as her beating gills futilely try to extract oxygen from air.

If you are a fish caught on a longline hook, you languish impaled for hours, sometimes days, before being dragged a mile or more to the boat deck. There, if not already dead, you usually succumb to suffocation. You may also sustain bites from predators from which, needless to say, you are helpless to escape.

Deeper-dwelling fishes face another peril: decompression. Decompression wreaks havoc on fishes because their gas-filled swim bladders—which function in buoyancy control—expand when the fish is hoisted toward the surface. As the swim bladder inflates, it presses against neighboring organs, which can cause their collapse and failure. More than a dozen studies published between 1964 and 2011 document lethal or sublethal injuries to commercially or recreationally fished species due to decompression. It's a nauseating list: esophageal eversion (the esophagus turns inside out and comes out of the mouth), exophthalmia (bulging of the eye from its orbit), arterial embolism (a sudden interruption of blood flow due to blockage by gas bubbles), kidney embolism, hemorrhage, organ torsion, damaged or displaced organs surrounding the swim bladder, and cloacal prolapse—a fish version of one's rectum turning inside out and emerging from the body.

Fishes raised in captivity do not have to die from decompression, crushing, or on a hook, but they don't exactly have it lucky. A 2002 review of fish slaughter studies concluded that the degree to which fishes suffer is "very high" when they are bled out (usually by slicing through the gills with a sharp knife), decapitated, put in a salt or ammonia bath (banned in Germany as inhumane for killing eels since 1999), or electrocuted. Suffocation, suffocation on ice, carbon dioxide narcosis, and anoxic water bath were classified as causing less but still "high" suffering. Some of these methods can cause immobility before loss of sensibility, leading to the illusion that suffering has ceased when it has not. Dying on ice is deemed welfare-unfriendly for fishes because it prolongs the process of suffocation. At room temperature, it takes an adult salmon about two and a half minutes to lose consciousness, and eleven minutes before all movement stops, whereas at near-freezing temperatures it takes much longer: more than nine minutes and more than three hours, respectively.

Collateral Damage

If killing captive fishes is little better than killing wild ones, at least fish farmers know who they're taking. In the wild, fishermen do not catch only what they are targeting; nets and hooks are indifferent to whom they catch. Unwanted fishes and other animals caught incidentally in the pursuit of the targeted species are referred to as bycatch. In commercial fishing, bycatch includes all seven kinds of sea turtles; dozens of seabirds, including albatrosses, gannets, shearwaters, razorbills, and petrels; practically every species of dolphin and whale; countless invertebrates; living corals; and, of course, a huge range of fish species. Because they are unwanted, they are typically thrown away.

Bycatch is common—very common. Estimates of how many sea creatures we toss back as unwanted waste vary, but they are consistently eye-popping. Try to visualize a pile of marine creatures weighing 200 million pounds, most of them dead, almost all doomed to die. That's the *daily* bycatch we reap from the seas.

According to the Fisheries and Aquaculture Department of the Food and Agriculture Organization (FAO), global yearly bycatch rates have been dropping, from around 29 million tons in the 1980s down to 7 million tons by 2001. Some credit for this might go to more selective fishing gear, and improved rules intended to reduce bycatch. But the trend is deceptive. The estimates from 1994 and 2005 that appear to show a decline cannot be reliably compared because they were calculated so differently. And, as targeted species have declined, fishermen have simply kept more of what used to be tossed overboard. Lower-value creatures, formerly discarded as trash, are being kept for human or animal food. It's for this reason that a quartet of wildlife analysts, chiefly with World Wildlife Fund International, proposed expanding the definition of bycatch to include "unmanaged" catch: nontargeted creatures nevertheless kept but for which no sustainable management plan exists. By this definition, bycatch today makes up 40 percent of the total global fish catch.

Some fisheries are more wasteful than others. Most notorious for their bycatch rates are shrimp fisheries. Because shrimps scuttle about on the bottom, catching them involves running those trawls we met earlier. Unwanted-fish-to-shrimp weight ratios average 1:1 to 3:1 for southeastern U.S. shrimp fisheries. Overall, 105 fish species have been recorded as bycatch on U.S. shrimp trawlers.

Bycatch has an insidious cousin: ghost nets. Fishing fleets abandon or lose many untold miles of synthetic fiber drift nets and bottom-set gill nets every year—about 640,000 tons of derelict gear in total, according to a recent analysis by World Animal Protection. These phantom menaces float beyond the grasp of human greed and continue entangling animals. The primary victims—dolphins, seals, sea birds, and sea turtles—become bait for other marine life, some of whom also become ensnared, until finally, by the sheer weight of all the dead bodies, the nets sink to the bottom of the sea.

Are we doing anything about the bycatch and ghost fishing debacles? Yes, and there has been some progress. Passage of the Marine Mammal Protection Act in 1972 helped cut yearly dolphin mortality in the U.S. tuna fishery from about half a million to 20,000. Further measures eventually brought the dolphin kill rate down to an estimated 3,000 per year by the mid-1990s. But dolphin populations have not recovered, and this is just one fishery. Worldwide, some 300,000 small whales, dolphins, and porpoises still die from entanglement in fishing nets each year, making this the leading killer of small cetaceans.

The situation is similar with seabirds. Baited longlines and wire struts called warp lines on trawlers had been killing about 100,000 albatrosses and petrels a year. Then, in 2008, the Albatross Task Force, a U.K. charity, showed in a pilot test off South Africa that the simple tactic of tying reusable pink strips—whose flapping motion has a scarecrow effect—to the lines and wires (at a cost of about twenty-two dollars per ship) could cut the toll by 85 percent. Under a multilateral agreement to protect pelagic seabirds, such simple bird-scaring designs are now recommended for implementation industry-wide. But albatrosses remain in big trouble, with seventeen of the twenty-two species considered vulnerable, endangered, or critically endangered, and the remaining five classified as "near threatened" by the International Union for Conservation of Nature.

In words often attributed to Joseph Stalin: "A single death is a tragedy; a million deaths is a statistic." When we are confronted with the astronomical numbers of animals who fall victim to our ocean plundering, we struggle to make an emotional connection to them. But if we were to interact with any one of those dolphins, those albatrosses, or for that matter any of the anonymous fishes hauled up to their deaths, we would come to know them as individuals. They would become someones, not somethings.

Finned

There are other ways to waste life at sea. Enter the world of shark finning. The practice involves catching sharks and cutting off their fins and tails to be used in shark fin soup, prized as a delicacy in China and other parts of Asia.

Shark finning is as brutal as it is profitable. Because handling large, muscular animals with sharp teeth on a slippery boat deck is dangerous work, killing them adds another layer of peril. So in the interest of speed and "efficiency," fishermen routinely slice the fins off the shark and throw the still-living animal (referred to as a "log") overboard to die of blood loss, suffocation, or compression as it slowly sinks into the abyss.

Iris Ho, with Humane Society International in Washington, D.C., is one of a growing cadre of campaigners working to end the shark fin trade. Having grown up in Taiwan, Ho has firsthand experience with shark fin soup from the days before she turned to animal protection. For centuries, shark fins were a rare extravagance reserved mainly for emperors, and it was not until the 1960s that advances in catching technologies made shark fins available to a broader class of consumers. By 2011, somewhere between 26 million and 73 million sharks were being slaughtered for their fins each year.

In an era of expanding animal and ocean advocacy stoked by the rapid dissemination of information via the Internet, ending shark finning has become a cause célèbre. A celebrity-driven campaign by the charity WildAid features the likes of Jackie Chan, David Beckham, and the basketball star Yao Ming. Revered in his native China, Yao appears in public service announcements, turning away shark fin soup when it is offered at a restaurant and urging others to do the same. Humane Society International has focused on community engagement campaigns, and momentum has grown. Chinese students designed campaigns to raise community awareness. One Walmart store in a major city of China showed shark movies on TV screens in its store, and sponsored a "no shark fin" pledge. As part of an austerity campaign against wasteful spending, the Chinese government issued a policy pronouncement against serving shark fins at official functions.

These campaigns are working. WildAid reports that 85 percent of surveyed Chinese consumers have given up shark fin soup within the past three years. As of late 2014, shark fin sales had dropped 82 percent in Guangzhou, which has replaced Hong Kong as China's shark fin trading hub, and retail and wholesale prices there had dropped by 47 and 57 percent, respectively, in two years. Dozens of commercial airlines have stopped shipping shark fins, and high-end hotel chains have removed shark fin dishes from their menus.

It remains to be seen how sharks will weather what surely must be the greatest assault ever to afflict their kind since their first ancestors appeared 450 million years ago. A shark's fins are not the only source of his woes. The trade in his meat has grown by 42 percent since 2000, totaling more than 258 million pounds. Despite having banned the finning of sharks at sea, the United States exported nearly 84,000 pounds of shark fins in 2011. How ironic it is that we have viewed sharks as fearsome killers when the ratio of their toll on us to our toll on them is about one to five million. Small wonder that some shark researchers are pursuing studies aimed at ending shark fishing.

Gone Fishin'

Commercial fishing, aquaculture, bycatch, and finning are all elements of fishing for monetary gain. What sort of impact might *recreational* fishing be having on fishes? The United States Fish and Wildlife Service describes recreational fishing—also known as angling or sportfishing—as one of the nation's most popular outdoor activities, attracting 33.1 million individuals sixteen years old and older in 2011. Worldwide, more than one in ten humans engages regularly in recreational fishing. Peer inside a sportfishing magazine—of which at least thirty are currently published in the United States—and you'll quickly see that recreational fishing is big business. In 2013, the American Sportfishing Association estimated that America's anglers spent $46 billion on fishing equipment, transportation, lodging, and other associated expenses.

While the unsustainability and cruelty of commercial fishing are increasingly recognized, recreational fishing retains a benign and beloved place in our culture. To wit, fishing scenes frequently appear in advertisements for pharmaceuticals and retirement communities—goods that have nothing directly to do with fishing.

Is angling really so benign? I doubt the fishes think so. Being hooked through the mouth (or worse) and forcibly transferred into an environment that causes suffocation certainly doesn't sound like something any of us would choose for an afternoon of peaceful recreation. If you've ever tried to remove a standard, barbed hook from a fish's mouth, you'll know that the barb is there for a reason, and the reason is not to make life easier for the fish. That little spur can cause damage to a fish's facial tissues even when it is removed carefully, and especially if it is forcefully torn out. I can still remember the dogged resistance posed by hooks, and the crackling sound they made as I worked them out with my inexperienced hands during my brief childhood fishing career. It is mostly up to chance which part of a fish's face gets impaled when a fisherman yanks the line upon detecting a nibble. Eye damage from hooks is surprisingly common, earning mention in a number of fishing studies. In a study of stream salmon, one in ten landed fishes sustained eye damage severe enough to be deemed likely to cause prolonged or permanent impaired vision.

Nowadays, anglers have the option of using barbless hooks, which can either be purchased or rendered that way with a pair of pliers. Barbless hooks probably originated in the United Kingdom, where catch-and-release fishing has been practiced for more than a century to prevent target species from disappearing in heavily fished waters. Hook extraction is easier with barbless hooks, and it can often be done without removing the fish from the water.

Hooks are not the only source of death and injury to recreationally caught fishes. The handling required to manage an unhappy wild creature is often necessarily rough. The protective, slimy layer of mucus surrounding the scales may be damaged by hands, landing nets, and hook extraction tools, leaving the fish more vulnerable to disease. Landing nets cause injuries ranging from severe fin abrasion to loss of scales and mucus, resulting in death rates from 4 percent to 14 percent. Pathogens also lurk. In a study of 242 largemouth basses caught in fishing tournaments and kept in submerged cages for four days of observation, four species of virulent bacteria were found in forty-two of seventy-six fishes with damaged skin. Another 8 percent had died before they were weighed at capture, and a further 25 percent died during the holding period, bringing the total death rate to one in three and suggesting that at least some of the infections were lethal.

Lastly, one might think that recreational fishing does not involve the decompression injuries sustained by fishes caught in deep-sea commercial fishing. In fact, some recreationally caught fishes are at sufficient depths to cause decompression injuries during the fish's forced transit to the surface. Nevertheless, a fish usually survives if quickly returned to the depths, and tools for doing this include a weighted crate that can be lowered and opened with a rope, and a commercial "fish descender."

Eaten

Whether the catch is commercial or recreational, when we eat fishes we are usually eating wildlife. Because humans prefer the taste of larger predatory fishes such as tunas, groupers, swordfishes, and mackerels, fisheries have tended to target them. During the twentieth century, humans have reduced the biomass of predatory fishes by more than two-thirds, and most of this alarming decline has occurred since the 1970s. Sylvia Earle puts it this way: "Think of everything in the fish market as bush meat. These are the eagles, the owls, the lions, the tigers, the snow leopards, the rhinoceroses of the ocean."

Perhaps no fish better exemplifies our consumption of wild predators than tunas. Eating tunas is like eating tigers. Like tigers, tunas are charismatic apex predators. And like tigers, tunas are big. The largest Atlantic bluefin tunas outsize the largest tiger, measuring nearly ten feet and weighing 1,500 pounds. Bristling with muscle and shaped like a bullet, a speeding tuna is as fast as an ambushing tiger. Lying at the top of their food chain, tunas require a lot of energy to fuel the growth and maintenance of their bodies. A tuna consumes her body weight in prey animals (mostly fishes, also squids and a few crustaceans) about every ten days. No thanks to the stacks of canned tuna grinning on grocery store shelves, most of the commercially fished species are in trouble. Atlantic and Pacific bluefin tunas are especially endangered, with populations estimated to be down by 85 percent and 96 percent, respectively, since 1960.

One of the dilemmas of approaching extinction is that, as you become rarer, you become more precious, which in turn makes you more valuable as a commodity. Today, one bluefin tuna can sell for over a million dollars. Ounce for ounce, that's twice the price of silver, and a huge incentive for a commercial fisherman.

Besides wildlife, there are other things we are eating when we eat a fish. Fish flesh is the most contaminated of all foods. Water flows downstream. Effluents find their way into organisms at the base of food chains, which in turn get concentrated through bioaccumulation as they move up food chains, ending up in the tissues of apex predators. Of 125,000 new chemicals developed since the industrial revolution, 85,000 have been found in fishes. It is well established that certain human demographics—most notably pregnant and nursing women and young children—are advised to limit consumption of fishes to avoid risk of exposure to mercury and other harmful chemicals. According to the physician Michael Greger, M.D., the author of *How Not to Die* and host of the popular website NutritionFacts.org, fish consumption is a leading source of mercury, dioxins, neurotoxins, arsenic, DDT, putrescine, AGEs, PCBs, PDBEs, and prescription drugs. Among the undesirable effects these contaminants can have on us are lowered intelligence; lower sperm counts; more symptoms of depression, anxiety, and stress; and earlier puberty.

So far, none of this is affecting policies or behavior. On the contrary, for years now, people in developed countries have been encouraged to increase their intakes of fatty fish by at least two- to threefold. The main problem with this advice—besides the fact that there are safer sources of omega-3 fatty acids than fish (for example, flaxseeds and walnuts)—is that it ignores the fact that feeding fishes to humans is unsustainable even at current levels of consumption.

This is not just an environmental problem; it is a geographic one. The combined effect of rising demand for fish and the collapse of fisheries is leading developed countries who can afford it, such as the United States, Japan, and European Union members, to increase imports from developing countries. The added pressure placed on these countries' coastal fishing grounds deprives local people of an important protein source for the sake of the developed world, whose major problems include overnutrition and physical inactivity.

Having witnessed the sharp decline in populations of fishes during her lifetime, Earle made a personal decision to stop eating them. "Ask yourself this," she says. "Is it more important to you to consume fish, or to think of them as being here for a larger purpose?"

Whether what we catch is intentional or incidental, our toll on marine life is huge. A 2015 joint study from the WWF and the Zoological Society of London concluded that fish populations halved between 1970 and 2012. Populations of some commercially heavily exploited species, including a group encompassing tuna, mackerel, and bonito, had fallen by almost 75 percent.

It is easy to condemn the cruelty and waste rampant in the commercial fishing industries. But consumers must acknowledge their complicity. In any supply-and-demand economy, demand is the fuel that drives the engine of supply. When we eat fishes, we fund their capture.

Is there any good news for fishes? Yes. Over the past quarter century, we have begun to take an unprecedented interest in animals as subjects of moral and ecological concern, and fishes are finally being swept into the current. "If an animal is sentient then it should be included in the moral circle," say five authors from veterinary, theological, and philosophical disciplines in a 2007 article on the ethics of fish-farming. Based on the clear evidence that they can feel pain, we may conclude that fishes should be given the benefit of the doubt.

Epilogue

The arc of the moral universe is long, but it bends toward justice.
—Martin Luther King, Jr.

Knowledge is a powerful thing; it informs ethics and fuels revolutions—witness the end of colonialism and institutionalized slavery and the advancement of women's and civil rights. These were triumphs of reason stoked by a growing sense of moral revulsion. Injustices, be they driven by greed, narrow-mindedness, prejudice, or all three, wither in the face of informed reason. The color of one's skin, one's religion, or having a womb or other arbitrary traits are simply no grounds for exploitation.

What about the number of legs, or having fins? The latter portion of the twentieth century saw unprecedented advances in concern for animals, including the rise of an increasingly sophisticated and effective animal rights movement. These trends continue to accelerate in the twenty-first century. The Humane Society of the United States, the world's most influential animal protection organization, reports that more than a thousand animal protection laws have been enacted in the United States since 2004—a number that rivals all of the animal protection laws enacted in

American history prior to 2000. In 1985, animal cruelty was a felony crime in just four U.S. states; by 2014, all fifty states had enacted such legislation. The public outcry over the shooting of an iconic African lion named Cecil by an American dentist in July 2015 illustrates the growing sympathy for animals' plight. Within a week, Cecil had become a household name, and close to 1.2 million people signed an online "Justice for Cecil" petition.

But a lion has a lot more charisma than a lionfish. I believe that the main source of our prejudices against fishes is their failure to show expressions that we associate with having feelings. "Fish are always in another element, silent and unsmiling, legless and dead-eyed," writes Jonathan Safran Foer in *Eating Animals*. In those flat, glassy eyes we struggle to see anything more than a vacant stare. We hear no screams and see no tears when their mouths are impaled and their bodies pulled from the water. Their unblinking eyes—constantly bathed in water and thus in no need of lids—amplify the illusion that they feel nothing. With a deficit of stimuli that normally trigger our sympathy, we are thus numbed to the fish's plight.

What we fail to account for when our sympathies falter is that the creature we are regarding is out of its element. Crying out in pain is as ineffective for a fish in air as crying out in pain is for us when we are submerged. Fishes are rigged to function, communicate, and express themselves underwater. Many do vocalize when they are hurt, but the sounds they produce evolved to pass through water, and rarely do we detect them. Even when we can notice signs of distress—flipping, thrashing, gills opening and closing as the animal tries in vain to take in oxygen—if we are schooled in the belief that they are just reflexive, we may shrug it off as nothing to be concerned about.

Today, we know far more about fishes than we did a century ago, even though what we know is a tiny fraction of what they know. Of the 30,000-plus fish species so far described, only a few hundred have been studied in any detail. The ones you have read about in this book are like celebrities among fishes. The most studied of them all, the zebrafish, *Danio rerio*—the "laboratory rat" of fishes—has been the subject of more than 25,000 published scientific papers, including more than 2,000 in 2015. (Not that we should envy them, as many of these studies are inhumane.) This serves to illustrate the boundless depths of inquiry and discovery to which any one of those 30,000 fish species could theoretically be subjected.

The previous section focused mainly on the way we exploit and abuse fishes. But our relationship to fishes is certainly not universally bad, and as our knowledge expands we are becoming less indifferent and more concerned for their welfare. When I did an informal online search for "fish welfare," the database IngentaConnect yielded seventy-one hits, sixty-nine of which were published since 2002. During the years I researched this book, I heard from dozens of correspondents who adore fishes and would never seek to harm them.

What many of these people like about fishes is not that they are like us. What is beautiful about them, and equally worthy of respect, is how they are *not* like us. Their different ways of being in the world are a source of fascination and admiration, and cause for sympathy. We can connect across the great divide that separates us, as when I have felt the gentle tugs of discus fishes rising to pluck food from my fingertips, or when a grouper fish approaches a trusted diver to receive caresses.

Among other things, fishes use their brains to survive and flourish, and one of the ways I have sought to raise the status of fishes has been to draw attention to their awareness and cognitive skill. But extolling the mental virtues of other species inflates the importance of intelligence, when intelligence really has little to do with moral standing. We do not deny basic moral rights to persons with developmental disabilities. Sentience—the capacity to feel, to suffer pain, to experience joy—is the bedrock of ethics. It is what qualifies one for the moral community.

Moral progress is good, and it is happening. Despite what we find in headlines, rates of human violence have declined significantly compared with historic levels. In his sweeping book *The Better Angels of Our Nature*, the psychologist Steven Pinker outlines a family of civilizing processes to explain this trend. Among them: the rise of democracies, the empowerment of women, the expansion of literacy, the formation of a global community, and the advance of reason. Today, new ideas spread almost instantaneously and seamlessly to all corners of the globe. Kickstarter campaigns generate funding groundswells for socially progressive projects, and independent foundations help get new ideas airborne.

Ever since we developed the concept of law, animals have been considered the legal property of humans. Yet even this fundamental paradigm—so deeply enshrined in our anthropocentric consciousness—is starting to shift. Local ordinances have changed animals' legal status from "property" to "companion" in at least eighteen U.S. cities since 2000. Depending on where you live (and who with), you might be one of more than 6 million Americans and Canadians officially recognized as "animal guardians." In May 2015, a New York supreme court judge held a hearing for two chimpanzees—used for years in invasive experiments at Stony Brook University—to have their rights defended by human attorneys, for unlawful imprisonment. Lawyers with the Nonhuman Rights Project are preparing to file further lawsuits on behalf of other animals.

Through laws, policy, and action, fishes are beginning to take their place in the moral community. In parts of Europe it is now unlawful to keep a goldfish—a naturally social animal who can live for decades—alone in a barren fishbowl. A law enacted in April 2008 by the Swiss federal parliament requires anglers to complete a course on catching fishes more humanely. The Dutch government has clearly stated the need for better stunning and killing methods for fishes, and the Foundation for the Protection of Fish has begun lobbying to turn those words into action. In

Germany, a 2013 law requires that all fishes be rendered unconscious before slaughter, and fishing tournaments where the catch is weighed before being thrown back in the water have been banned, along with the use of minnows as live bait. In Norway, the use of carbon dioxide stunning was banned as inhumane in 2010.

Beyond laws for fishes, there is passion. For many of us, fishes inspire not just concern, but love. In researching this book I received letters from people who truly adored their fishes. A college professor from Spokane, Washington, wrote that she had grown to love a goldfish she rescued from being flushed; Pearl greeted her daily, swimming to the surface to feed from her hand, and when Pearl died at age seventeen, she described the loss as "like losing a beloved family cat or dog." A professional from Gainesville, Florida, developed a game with her blue discus fish, Jasper, in which they chased each other from each side of the glass; she told me that "when I would cup my hands just below the waterline, he would turn on his side and swim up into my hands and just lie there while I stroked his side." A businesswoman from Portland, Oregon, shared this description of her ten-year-old fahaka pufferfish, Mango—a close relative of the fish on the cover of this book:

> I've had him his whole life (nine years and counting) and he's not much different than my dog in that he wiggles uncontrollably when I get home and is extremely affectionate and interactive with me. We often get into these staring contests and he usually wins. I love this fish like I've never loved a fish. Most everyone I know has met Mango, and they're mesmerized by him. I'm certain Mango has altered people's perceptions of fishes.

And there are those who will simply go the extra mile for a fish. I have such a friend, who received an anonymous call, went to the address involved, and negotiated the rescue of three large koi fishes from a filthy, putrid tank where they had languished for eleven years. She drove them two hours to a well-kept koi pond at an Asian restaurant, where they now live comfortably in the company of their kind.

That rescue is just one of a growing number of acts of kindness toward fishes. One need only search that modern conduit of the amateur videographer, YouTube, to find scenes of divers extracting hooks from the mouths of sharks, or cutting fishing lines and nets from the fins of manta rays; beachcombers rescuing stranded fishes, and people with buckets transferring fishes from drying rivers and lake beds. I have an ichthyologist friend, a retired biology professor who, growing tired of killing fishes on teaching and collecting expeditions, invented a portable device that allows aquatic animals captured in the field to be photographed and released on-site. Sales of his teaching-photographic tank have saved over a million fishes from being preserved in formaldehyde to sit mostly unused on museum shelves. Another biologist has founded Fish Feel, North America's first organization dedicated to the protection of our underwater cousins. And you may not know that the Sea Shepherd Conservation Society, protagonists of the popular *Whale Wars* television series, also campaigns to save salmons, cods, bluefin tunas, Antarctic tooth fishes, and sharks. Not one to mince words, Sea Shepherd's founder, Paul Watson, told me, "When I see a salmon farm, I see slavery and the debasement of the spirit of the fish that West Coast First Nations viewed as the buffalo of the sea . . . One of my greatest moments of satisfaction was cutting the nets of a Maltese poacher off the coast of Libya and releasing eight hundred bluefin tuna. They sped through that opening like thoroughbred racehorses."

With the rise of reason, and a growing awareness of our interdependence with all life, humanity is on a course toward a more inclusive, more enlightened era. Basic principles of respect for all members of our own kind are gradually being extended to other beings once excluded.

But for now, for every fish saved we continue to kill so many. As I write this, news reached me of 75,000 dead menhadens washing up on a beach along Virginia's eastern shore, after a fishing net tore. Photos of their gaping, rotting bodies stretching to the horizon remind me that our subject's name is synonymous with its own demise, for the word "fish" means both the animal and the act of catching it.

Let me finish with a story that brought tears to my eyes when I first read it. The woman who shared it with me believes she was three at the time it happened, and it is her earliest memory. There were three small fishes in the house, living out their lives in an aquarium atop the fireplace mantel. This, she was later told, "was so that they might be up high and safe from little forces of energy that like to play and climb and run about." The little girl had also been taught that one must be careful of water, as we can't breathe in it. With her limited knowledge of the laws of nature at her tender age, she reasoned that the fishes couldn't breathe in there either. For weeks she worried about those fishes slowly drowning in the tank on the mantel. She felt a personal responsibility to rescue them.

One day when the family was leaving the house, she made sure she was the last to go. When everyone was out the door and the coast was clear, she climbed the mantel with the help of some chairs and nearby cupboards to carry out her rescue. She didn't have a plan beyond freeing the fishes from their watery grave. She also doesn't think she understood death or what happens when one drowns, except that it was painful, like getting water up your nose in the bathtub. There was a small net to remove debris from the tank, so she used it to lift the fishes out, placing them on the mantel. A parent returned to usher her out, and she left.

She doesn't remember the fishes' fate, but she didn't see them again after that. She thought about them often throughout kindergarten, keeping the recollection vivid among blurry memories. The years have not dulled her deep early empathy for animals. To this day, forty years later, it troubles her that she wanted to save someone but instead made them suffer.

This story resonates with several themes in this book. An innocent young child's misguided belief that fishes, like us, need to breathe air to survive represents our collective ignorance of fishes. Her act of removing them from their element to suffocate symbolizes the suffering they endure at our hands (even though her intention was a world away from the common belief that their role on Earth is to be our food and recreation). And her extraordinary sympathy, expressed at such a tender age and still felt today, reminds us of the infinite potential for humans, when we are aware, to do good in the world.

Notes

PROLOGUE

7 *rarely presented as a number of individuals*: FAO (Food and Agriculture Organization of the United Nations), *The State of World Fisheries and Aquaculture 2012* (Rome, Italy: Fisheries and Aquaculture Department, FAO, 2012).

7 *Stephen Cooke and Ian Cowx . . . estimated in 2004*: Stephen J. Cooke and Ian G. Cowx, 2004. "The Role of Recreational Fisheries in Global Fish Crises," *BioScience* 54 (2004): 857–59.

7 *47 billion fishes were being landed* recreationally: Steven J. Cooke and Ian G. Cowx, "The Role of Recreational Fisheries in Global Fish Crises," *BioScience* 54, no. 9 (2004): 857–59. This is an admittedly rather crude estimate, based on extrapolation from rates of recreational fishing catch from Canada to the global human population.

7 *One study reports*: Daniel Pauly and Dirk Zeller, "Catch Reconstructions Reveal That Global Marine Fisheries Catches Are Higher than Reported and Declining." *Nature Communications* 7 (2016):10244 doi: 10.1038/ncomms10244.

7 *The leading causes of death*: D. H. F. Robb and S. C. Kestin, "Methods Used to Kill Fish: Field Observations and Literature Reviewed," *Animal Welfare* 11, no. 3 (2002): 269–82.

8 *In the immortal words*: Widely attributed to Anthony de Mello (1931–1987), Jesuit priest and inspirational speaker/writer. See www.beyondpoetry.com/anthony-de-mello.html (and many other sources).

PART I: THE MISUNDERSTOOD FISH

9 *"We shall not cease from exploration"*: T. S. Eliot, "Little Gidding" (1942), in *Four Quartets* (New York: Harcourt Brace, 1943).

11 *According to FishBase—the largest*: Rainer Froese and Alexander Proelss, "Rebuilding Fish Stocks No Later Than 2015: Will Europe Meet the Deadline?" *Fish and Fisheries* 11, no. 2 (2010): 194–202.

11 *When we refer to "fish"*: Colin Allen, "Fish Cognition and Consciousness," *Journal of Agricultural and Environmental Ethics* 26, no. 1 (2013): 25–39.

11 *Members of both groups*: Gene Helfman, Bruce B. Collette, and Douglas E. Facey, *The Diversity of Fishes* (Oxford, UK: Blackwell, 1997).

12 *A third distinct group of fishes*: Gene Helfman and Bruce B. Collette, *Fishes: The Animal Answer Guide* (Baltimore: The Johns Hopkins University Press, 2011).

12 *A tuna is actually more closely related*: Allen, "Fish Cognition and Consciousness."

12 *As the author Sy Montgomery notes*: Sy Montgomery, "Deep Intellect: Inside the Mind of the Octopus," *Orion*, November/December 2011.

13 *We might also think of jaws*: Donald R. Prothero, *Evolution: What the Fossils Say and Why It Matters* (New York: Columbia University Press, 2007).

14 *About this discovery and John Long*: The relevant segment of Attenborough's lecture can be viewed at: www.youtube.com/watch?v=OXqgFkeTnJI.

15 *According to the National Oceanic and Atmospheric Administration*: National Geographic, *Creatures of the Deep Ocean* [documentary film], 2010.

15 *The deep sea is the largest habitat*: Xabier Irigoien et al., "Large Mesopelagic Fishes Biomass and Trophic Efficiency in the Open Ocean," *Nature Communications* 5 (2014): 3271.

15 *It's a shallow idea*: Tony Koslow, *The Silent Deep: The Discovery, Ecology, and Conservation of the Deep Sea* (Chicago: University of Chicago Press, 2007), 48.

15 *Technology is just beginning to afford us*: Helfman, Collette, and Facey, *Diversity of Fishes* (1997).

15 *Between 1997 and 2007, 279 new species*: David Alderton, "Many Fish Identified in the Past Decade," FishChannel.com, December 24, 2008, www.fishchannel.com/fish-news/2008/12/24/mekong-fish-discoveries.aspx.

15 *Given the current rate, experts predict*: Allen, "*Fish Cognition and Consciousness.*"

16 *The smallest fish*: *Pandaka pygmaea* has some competition: www.scholastic.com/browse/article.jsp?id=11044; http://en.microcosmaquariumexplorer.com/wiki/Fish_Facts_-_Smallest_Species. Here's a blog post that jocularly mourns that it has been "out-smalled": http://unholyhours.blogspot.com/2006/01/farewell-to-pandaka-pygmaea.html.

16 *Adult* Pandaka pygmaea *are only*: John R. Norman and Peter H. Greenwood, *A History of Fishes*, 3rd rev. ed. (London: Ernest Benn Ltd., 1975).

16 *At less than half an inch*: Tierney Thys, "For the Love of Fishes," in *Oceans: The Threats to Our Seas and What You Can Do to Turn the Tide*, ed. Jon Bowermaster (New York: Public Affairs, 2010), 137–42.

16 *It is estimated that female deep-sea anglerfishes*: Gene Helfman, Bruce B. Collette, Douglas E. Facey, and Brian W. Bowen, *The Diversity of Fishes: Biology, Evolution, and Ecology*, 2nd ed. (Chichester, UK: Wiley-Blackwell, 2009).

16 *At the time that Peter Greenwood*: Norman and Greenwood, *History of Fishes*.

17 *A single ling, five feet long*: Norman and Greenwood.

17 *But the growth champion among vertebrates*: E. W. Gudger, "From Atom to Colossus," *Natural History* 38 (1936): 26–30.

18 *a heavily fished species that you might have dissected*: Mark W. Saunders and Gordon A. McFarlane, "Age and Length at Maturity of the Female Spiny Dogfish, *Squalus acanthias*, in the Strait of Georgia, British Columbia, Canada," *Environmental Biology of Fishes* 38, no. 1 (1993): 49–57.

18 *Sharks have a placental structure*: Helfman, Collette, and Facey, *Diversity of Fishes* (1997).

18 *Frilled sharks carry their babies*: Helfman et al., 1997.

18 *Once airborne, the lower lobe*: Norman and Greenwood, *History of Fishes*.

19 *I do not see why having*: As Rod Preece and Lorna Chamberlain say in their 1993 book, *Animal Welfare and Human Values*: "We can find no justification for that prevalent belief . . . that cold-blooded animals are . . . less sentient than warm-blooded animals." Vladimir Dinets, a Russian American scientist who has traveled the world observing wild crocodiles and revealing such surprises as tool use, coordinated hunting, courtship parties, and tree climbing, is more blunt: "Most humans are warm-blooded bigots" (Vladimir Dinets, personal communication, March 18, 2014).

20 *Tunas, swordfishes, and some sharks*: Helfman et al., 1997.

20 *They achieve this by capturing heat*: Francis G. Carey and Kenneth D. Lawson, "Temperature Regulation in Free-Swimming Bluefin Tuna," *Comparative Physiology and Biochemistry Part A: Physiology* 44, no. 2 (1973): 375–92.

20 *many sharks have a large vein*: Nancy G. Wolf, Peter R. Swift, and Francis G. Carey, "Swimming Muscle Helps Warm the Brain of Lamnid Sharks," *Journal of Comparative Physiology B* 157 (1988): 709–15.

20 *The large, predatory billfishes*: Helfman et al., *Diversity of Fishes* (1997).

20 *the first truly endothermic fish*: Nicholas C. Wegner et al., "Whole-Body Endothermy in a Mesopelagic Fish, the Opah, *Lampris guttatus*," *Science* 348 (2015): 786–89.

21 *So about half of fish species*: Culum Brown, "Fish Intelligence, Sentience and Ethics," *Animal Cognition* 18, no. 1 (2015): 1–17.

21 *The Age of Teleosts*: Prothero, *Evolution: What the Fossils Say.*
22 *you will rarely see a stationary fish*: Norman and Greenwood, *History of Fishes.*

PART II: WHAT A FISH PERCEIVES
WHAT A FISH SEES

23 *"There is no truth"*: Gustave Flaubert, unsourced quote. Taken from https://en.wikiquote.org/wiki/Talk:Gustave_Flaubert.
25 *"red-gold, water-precious"*: D. H. Lawrence, "Fish" (1921), in *Birds, Beasts and Flowers: Poems* (London: Martin Secker, 1923).
26 *Like most vertebrate eyeballs*: Helfman et al., *Diversity of Fishes* (1997).
26 *With a spherical lens*: David McFarland, ed., *The Oxford Companion to Animal Behavior* (Oxford: Oxford University Press, 1982; reprint ed., 1987).
27 *Seahorses, blennies, gobies, and flounders*: Arthur A. Myrberg Jr. and Lee A. Fuiman, "The Sensory World of Coral Reef Fishes," in *Coral Reef Fishes: Dynamics and Diversity in a Complex Ecosystem*, ed. Peter F. Sale, 123–48 (Burlington, MA: Academic Press/Elsevier, 2002); Mark Sosin and John Clark, *Through the Fish's Eye: An Angler's Guide to Gamefish Behavior* (New York: Harper and Row, 1973).
27 *Although a team of scientists from Israel*: Ofir Avni et al., "Using Dynamic Optimization for Reproducing the Chameleon Visual System," presented at the 45th IEEE Conference on Decision and Control, San Diego, CA, December 13–15, 2006.
27 *The entire migration takes just five days*: Helfman et al., *Diversity of Fishes* (2009), 138.
28 *Flexible genetic coding*: David Alderton, "New Study Unveils Mysteries of Vision in *Anableps anableps*, the Four-Eyed Fish," FishChannel.com, July 25, 2011, www.fishchannel.com/fish-news/2011/07/25/anableps-four-eyedfish.aspx.
29 *Swordfishes can heat up their eyes*: Helfman et al., *Diversity of Fishes* (1997).
29 *The heat is generated by a countercurrent*: Kerstin A. Fritsches, Richard W. Brill, and Eric J. Warrant, "Warm Eyes Provide Superior Vision in Swordfishes," *Current Biology* 15, no. 1 (2005): 55–58.
30 *This enables a fish*: Sosin and Clark, *Through the Fish's Eye.*
30 *The refractive properties of calm water*: Sosin and Clark.
30 *discovered evidence of rods and cones*: Gengo Tanaka et al., "Mineralized Rods and Cones Suggest Colour Vision in a 300 Myr–Old Fossil Fish," *Nature Communications* 5 (2014): 5920; Sumit Passary, "Scientists Discover Rods and Cones in 300-Million-Year-Old Fish Eyes. What Findings Suggest," *Tech Times*, December 24, 2014, www.techtimes.com/articles/22888/20141224/scientists-discover-rods-and-cones-in-300-million-year-old-fish-eyes-what-findings-suggest.htm.

31 *most modern bony fishes are tetrachromatic*: Brown, "Fish Intelligence."

31 *This helps explain why about one hundred*: George S. Losey et al., "The UV Visual World of Fishes: A Review," *Journal of Fish Biology* 54, no. 5 (1999): 921–43.

31 *the value of having a wider visual spectrum*: Ulrike E. Siebeck et al., "A Species of Reef Fish That Uses Ultraviolet Patterns for Covert Face Recognition," *Current Biology* 20, no. 5 (2010): 407–10.

31 *Furthermore, because the predators of damselfishes*: Ulrike E. Siebeck and N. Justin Marshall, "Ocular Media Transmission of Coral Reef Fish—Can Coral Reef Fish See Ultraviolet Light?" *Vision Research* 41 (2001): 133–49.

32 *I remember flipping through a biology textbook*: Photo of flounder camouflaged on checkerboard: http://users.rcn.com/jkimball.ma.ultranet/BiologyPages/C/Chromatophores.html.

33 *Being seen is a high priority for a fish*: Interpretive sign at the Smithsonian National Museum of Natural History, Washington, D.C., September 2012.

33 *These glowing adornments enhance*: Norman and Greenwood, *History of Fishes.*

34 *Ponyfishes have a peculiar method*: D. J. Woodland et al., "A Synchronized Rhythmic Flashing Light Display by Schooling 'Leiognathus Splendens' (Leiognathidae: Perciformes)," *Marine and Freshwater Research* 53, no. 2 (2002): 159–62; Akara Sasaki et al., "Field Evidence for Bioluminescent Signaling in the Pony Fish, *Leiognathus elongatus*," *Environmental Biology of Fishes* 66 (2003): 307–11.

34 *Mated pairs of flashlight fishes*: James G. Morin et al., "Light for All Reasons: Versatility in the Behavioral Repertoire of the Flashlight Fish," *Science* 190 (1975): 74–76.

34 *Named for a capacious lower mandible*: Stephen R. Palumbi and Anthony R. Palumbi, *The Extreme Life of the Sea* (Princeton: Princeton University Press, 2014).

35 *Irene Pepperberg's touching memoir*: Irene Pepperberg, *Alex & Me: How a Scientist and a Parrot Uncovered a Hidden World of Animal Intelligence—and Formed a Deep Bond in the Process* (New York: HarperCollins, 2008), 202.

36 *in a captive study of redtail splitfins*: Valeria Anna Sovrano, Liliana Albertazzi, and Orsola Rosa Salva, "The Ebbinghaus Illusion in a Fish (*Xenotoca eiseni*)," *Animal Cognition* 18 (2015): 533–42.

36 *the more familiar Müller-Lyer illusion*: V. A. Sovrano, "Perception of the Ebbinghaus and Müller-Lyer Illusion in a Fish (*Xenotoca eiseni*)," poster presented at CogEvo 2014, the 4th Rovereto Workshop on Cognition and Evolution, Rovereto, Italy, July 7–9.

36 *Studies of goldfishes and bamboo sharks*: O. R. Salva, V. A. Sovrano, and Giorgio Vallortigara, "What Can Fish Brains Tell Us About Visual Perception?" *Frontiers in Neural Circuits* 8 (2014): 119, doi:10.3389/fncir.2014.00119.

38 *A further enhancement is having a tail end*: Desmond Morris, *Animalwatching: A Field Guide to Animal Behavior* (London: Jonathan Cape, 1990).

WHAT A FISH HEARS, SMELLS, AND TASTES

40 *"The universe is full of magical things"*: Eden Phillpotts, *A Shadow Passes* (London: Cecil Palmer and Hayward, 1918), 19. Often misattributed to W. B. Yeats or Bertrand Russell.

40 *Fishes have separate organs*: Helfman et al., *Diversity of Fishes* (1997); A. O. Kasumyan and Kjell B. Døving, "Taste Preferences in Fish," *Fish and Fisheries* 4, no. 4 (2003): 289–347.

40 *Despite the common assumption that fishes are silent*: Friedrich Ladich, "Sound Production and Acoustic Communication," in *The Senses of Fish: Adaptations for the Reception of Natural Stimuli*, Gerhard Von der Emde et al., eds., 210–30 (Dordrecht, Netherlands: Springer, 2004).

41 *the options of grating their teeth*: Norman and Greenwood, *History of Fishes*.

41 *hums, whistles, thumps, stridulations, creaks*: Arthur A. Myrberg Jr. and M. Lugli, "Reproductive Behavior and Acoustical Interactions," in *Communication in Fishes*, Vol. 1, ed. Friedrich Ladich et al., 149–76 (Enfield, NH: Science Publishers, 2006).

41 *It was only in the past century*: Helfman and Collette, *Fishes: The Animal Answer Guide*.

41 *Karl von Frisch (1886–1982)*: Tania Munz, "The Bee Battles: Karl von Frisch, Adrian Wenner and the Honey Bee Dance Language Controversy," *Journal of the History of Biology* 38, no. 3 (2005): 535–70.

42 *These bones have become separated*: Norman and Greenwood, *History of Fishes*.

42 *similar to the middle ear ossicles*: Norman and Greenwood, *History of Fishes*.

43 *well above the upper human limit*: David A. Mann, Zhongmin Lu, and Arthur N. Popper, "A Clupeid Fish Can Detect Ultrasound," *Nature* 389 (1997): 341; D. A. Mann et al., "Detection of Ultrasonic Tones and Simulated Dolphin Echolocation Clicks by a Teleost Fish, the American Shad (*Alosa sapidissima*)," *Journal of the Acoustical Society of America* 104, no. 1 (1998): 562–68.

43 *Sensitivity to infrasound*: O. Sand and H. E. Karlsen, "Detection of Infrasound and Linear Acceleration in Fishes," *Philosophical Transactions of the Royal Society of London B: Biological Sciences* 355 (2000): 1295–98.

43 *For instance, the delicate hair cells*: Robert D. McCauley, Jane Fewtrell, and Arthur N. Popper, "High Intensity Anthropogenic Sound Damages Fish Ears," *The Journal of the Acoustical Society of America* 113, no. 1 (2003): 638–42.

43 *Intense noise produced by seismic air-gun*: Arill Engås et al., "Effects of Seismic Shooting on Local Abundance and Catch Rates of Cod (*Gadus morhua*) and Haddock (*Melanogrammus aeglefinus*)," *Canadian Journal of Fisheries and Aquatic Sciences* 53 (1996): 2238–49.

43 *they are proficient at sound directionality*: Stéphan Reebs, *Fish Behavior in the Aquarium and in the Wild* (Ithaca, New York: Comstock Publishing Associates/Cornell University Press, 2001).

44 *This is why anglers sitting in a boat*: Sosin and Clark, *Through the Fish's Eye.*

44 *Fishermen along the Atlantic coast of Ghana*: Sosin and Clark. An essay by a Ghanaian fisherman that describes listening to fishes can also be found here: B. Konesni, *Songs of the Lalaworlor: Musical Labor on Ghana's Fishing Canoes,* June 14, 2008, www.worksongs.org/blog/2013/10/18/songs-of-the-lalaworlor-musical-labor-on-ghanas-fishing-canoes.

44 *A third sound occurs when a piranha*: Sandie Millot, Pierre Vandewalle, and Eric Parmentier, "Sound Production in Red-Bellied Piranhas (*Pygocentrus nattereri,* Kner): An Acoustical, Behavioural and Morphofunctional Study," *Journal of Experimental Biology* 214 (2011): 3613–18.

45 *Ava Chase, a research scientist at Harvard*: Ava R. Chase, "Music Discriminations by Carp (*Cyprinus carpio*)," *Animal Learning and Behavior* 29, no. 4 (2001): 336–53.

46 *"It appears that [koi] can discriminate"*: Chase, "Music Discriminations," 352.

46 *Despite their skill as music connoisseurs*: Richard R. Fay, "Perception of Spectrally and Temporally Complex Sounds by the Goldfish (*Carassius auratus*)," *Hearing Research* 89 (1995): 146–54.

47 *team from the Agricultural University of Athens*: Sofronios E. Papoutsoglou et al., "Common Carp (*Cyprinus carpio*) Response to Two Pieces of Music ("Eine Kleine Nachtmusik" and "Romanza") Combined with Light Intensity, Using Recirculating Water System," *Fish Physiology and Biochemistry* 36, no. 3 (2009): 539–54.

48 *A 2015 review of 70 clinical trials*: Jenny Hole et al., "Music as an Aid for Postoperative Recovery in Adults: A Systematic Review and Meta-Analysis," *Lancet* 386 (2015): 1659–71.

48 *"I am not at all convinced that music"*: Karakatsouli, personal communication, June 2015.

48 *what might loosely be termed* flatulent communication: Ben Wilson, Robert S. Batty, and Lawrence M. Dill, "Pacific and Atlantic Herring Produce Burst Pulse Sounds," *Proceedings of the Royal Society of London, B: Biological Sciences* 271, supplement 3 (2004): S95–S97.

49 *They use chemical cues*: Wilson et al., "Herring Produce Burst Pulse Sounds."

49 *Sticklebacks, for example, use smell*: Nicole E. Rafferty and Janette Wenrick Boughman, "Olfactory Mate Recognition in a Sympatric Species Pair of Three-Spined Sticklebacks," *Behavioral Ecology* 17, no. 6 (2006): 965–70.

49 *Unlike those of other vertebrates, fishes' nostrils*: Norman and Greenwood, *History of Fishes*.

49 *Some fishes expand and contract their nostrils*: Sosin and Clark, *Through the Fish's Eye*.

49 *Signals from the epithelium are sent*: Toshiaki J. Hara, "Olfaction in Fish," *Progress in Neurobiology* 5, part 4 (1975): 271–335.

50 *A sockeye salmon can sense shrimp extract*: Sosin and Clark, *Through the Fish's Eye*.

50 *Once again, we owe it to Karl von Frisch*: Karl von Frisch, "The Sense of Hearing in Fish," *Nature* 141 (1938): 8–11; "Über einen Schreckstoff der Fischhaut und seine biologische Bedeutung," *Zeitschrift für vergleichende Physiologie* 29, no. 1 (1942): 46–145.

50 *And it is potent stuff*: Reebs, *Fish Behavior*.

50 *Schreckstoff must have evolved long ago*: R. Jan F. Smith, "Alarm Signals in Fishes," *Reviews in Fish Biology and Fisheries* 2 (1992): 33–63; Wolfgang Pfeiffer, "The Distribution of Fright Reaction and Alarm Substance Cells in Fishes," *Copeia* 1977, no. 4 (1977): 653–65.

51 *When they smell the poop*: Grant E. Brown, Douglas P. Chivers, and R. Jan F. Smith,"Fathead Minnows Avoid Conspecific and Heterospecific Alarm Pheromones in the Faeces of Northern Pike," *Journal of Fish Biology* 47, no. 3 (1995): 387–93.; "Effects of Diet on Localized Defecation by Northern Pike, *Esox lucius*," *Journal of Chemical Ecology* 22, no. 3 (1996): 467–75.

51 *It is probably due to olfactory skills*: Brown, Chivers, and Smith, "Localized Defecation by Pike: A Response to Labelling by Cyprinid Alarm Pheromone?" *Behavioral Ecology and Sociobiology* 36 (1995): 105–10.

51 *Juvenile lemon sharks react to the odor*: Robert E. Hueter et al., "Sensory Biology of Elasmobranchs," in *Biology of Sharks and Their Relatives*, ed. Jeffrey C. Carrier, John A. Musick, and Michael R. Heithaus (Boca Raton, FL: CRC Press, 2004).

51 *Instead, one may just learn*: Laura Jayne Roberts and Carlos Garcia de Leaniz, "Something Smells Fishy: Predator-Naïve Salmon Use Diet Cues, Not Kairomones, to Recognize a Sympatric Mammalian Predator," *Animal Behaviour* 82, no. 4 (2011): 619–25.

52 *Experiments from the 1950s showed*: W. N. Tavolga, "Visual, Chemical and Sound Stimuli as Cues in the Sex Discriminatory Behaviour of the Gobiid Fish *Bathygobius soporator*," *Zoologica* 41 (1956): 49–64.

52 *Female sheepshead swordtails from Mexico*: Heidi S. Fisher and Gil G. Rosenthal, "Female Swordtail Fish Use Chemical Cues to Select Well-Fed Mates," *Animal Behaviour* 72 (2006): 721–25.

52 *Male deep-sea anglerfishes illustrate the interplay*: Theodore W. Pietsch, *Oceanic Anglerfishes: Extraordinary Diversity in the Deep Sea* (Berkeley, CA: University of California Press, 2009).

52 *The male anglerfish's nostrils*: Pietsch, *Oceanic Anglerfishes*.

53 *Consider a 2011 study of our friends*: Gil G. Rosenthal et al., "Tactical

Release of a Sexually-Selected Pheromone in a Swordtail Fish," *PLoS ONE* 6, no. 2 (2011): e16994, doi:10.1371/journal.pone.0016994.

53 *the primary organs of taste*: For an excellent review of taste preferences in fishes, see Kasumyan and Døving, "Taste Preferences in Fish."

53 *fishes are quite literally immersed*: McFarland, *Oxford Companion to Animal Behavior*; Sosin and Clark, *Through the Fish's Eye.*

54 *For instance, a fifteen-inch channel catfish*: Thomas E. Finger et al., "Postlarval Growth of the Peripheral Gustatory System in the Channel Catfish, *Ictalurus punctatus*," *The Journal of Comparative Neurology* 314, no. 1 (1991): 55–66.

54 *Cavefishes also benefit from a bonanza*: Yoshiyuki Yamamoto, "Cavefish," *Current Biology* 14, no. 22 (2004): R943.

54 *Many bottom-feeders, including catfishes*: Norman and Greenwood, *History of Fishes.*

54 *Stéphan Reebs, the author*: Reebs, *Fish Behavior*, 86.

NAVIGATION, TOUCH, AND BEYOND

56 *"When one flesh is waiting"*: Wallace Stegner, *Angle of Repose* (New York: Doubleday, 1971).

56 *Swordfishes, parrotfishes, and sockeye salmons*: Helfman et al., *Diversity of Fishes* (2009).

56 *Yet others can use dead reckoning*: Victoria A. Braithwaite and Theresa Burt De Perera, "Short-Range Orientation in Fish: How Fish Map Space," *Marine and Freshwater Behaviour and Physiology* 39, no. 1 (2006): 37–47.

57 *By isolating cells from the nasal passages*: Stephan H. K. Eder et al., "Magnetic Characterization of Isolated Candidate Vertebrate Magnetoreceptor Cells," *Proceedings of the National Academy of Sciences of the United States of America* 109 (2012): 12022–27.

57 *Years later, they retrace their paths*: Andrew H. Dittman and Thomas P. Quinn, "Homing in Pacific Salmon: Mechanisms and Ecological Basis," *Journal of Experimental Biology* 199 (1996): 83–91.

57 *In a less invasive*: Arthur D. Hasler and Allan T. Scholz, *Olfactory Imprinting and Homing in Salmon: Investigations into the Mechanism of the Homing Process* (Berlin: Springer-Verlag, 1983).

58 *Might a salmon also use vision*: Hiroshi Ueda et al., "Lacustrine Sockeye Salmon Return Straight to Their Natal Area from Open Water Using Both Visual and Olfactory Cues," *Chemical Senses* 23 (1998): 207–12.

59 *The lateral line is usually visible*: Norman and Greenwood, *History of Fishes.*

59 *fishes swimming in close proximity*: Myrberg and Fuiman, "Sensory World of Coral Reef Fishes."

59 *Blind cavefishes can form mental maps*: T. Burt de Perera, "Fish Can Encode Order in Their Spatial Map," *Proceedings of the Royal Society B: Biological Sciences* 271 (2004): 2131–34, doi:10.1098/rspb.2004.2867.

60 *visual and lateral line sensory systems operate independently*: T. Burt de Perera and V. A. Braithwaite, "Laterality in a Non-Visual Sensory Modality—The Lateral Line of Fish," *Current Biology* 15, no. 7 (2005): R241–R242.

60 *swimming fishes are only half as likely*: Brian Palmer, "Special Sensors Allow Fish to Dart Away from Potential Theats at the Last Moment," *Washington Post*, November 26, 2012, www.washingtonpost.com/national/health-science/special-sensors-allow-fish-to-dart-away-from-potential-theats-at-the-last-moment/2012/11/26/574d0960-3254-11e2-bb9b-288a310849ee_story.html.

60 *Electrical sensitivity is widespread in sharks*: Mark E. Nelson, "Electric Fish," *Current Biology* 21, no. 14 (2011): R528–R529.

61 *These pores are called the* ampullae of Lorenzini: R. Douglas Fields, "The Shark's Electric Sense," *Scientific American* 297 (2007): 74–81.

61 *The function of the ampullae of Lorenzini*: R. W. Murray, "Electrical Sensitivity of the Ampullae of Lorenzini," *Nature* 187 (1960): 957, doi:10.1038/187957a0.

61 *Such is the sensitivity of this system*: Helfman et al., *Diversity of Fishes* (1997).

61*n* *They have layers of fatty tissue*: Nelson, "Electric Fish."

62 *the fishes showed an "astounding" ability*: Stephen Paintner and Bernd Kramer, "Electrosensory Basis for Individual Recognition in a Weakly Electric, Mormyrid Fish, *Pollimyrus adspersus* (Günther, 1866)," *Behavioral Ecology & Sociobiology* 55 (2003): 197–208. doi:10.1007/s00265-003-0690-4.

62 *EODs also communicate social status*: Nelson, "Electric Fish."

62 *Dominant individuals may chase trespassers*: Andreas Scheffel and Bernd Kramer, "Intra- and Interspecific Communication among Sympatric Mormyrids in the Upper Zambezi River," in Ladich et al., eds., *Communication in Fishes*, 733–51 (Enfield, NH: Science Publishers, 2006).

63 *They deal with it*: Theodore H. Bullock, Robert H. Hamstra Jr., and Henning Scheich, "The Jamming Avoidance Response of High Frequency Electric Fish," *Journal of Comparative Physiology* 77, no. 1 (1972): 1–22.

63 *Fishes in a social group*: A. S. Feng, "Electric Organs and Electroreceptors," in *Comparative Animal Physiology*, 4th ed., ed. C. L. Prosser, 217–34 (New York: John Wiley and Sons, 1991).

63 *team up as shoaling partners*: Scheffel and Kramer, "Intra- and Interspecific Communication."

63 *Much of that gray matter*: Helfman et al., *Diversity of Fishes* (1997).

64 *In the evolutionary arms race*: Helfman et al., 1997.

65 *a video clip by a puzzled viewer*: www.youtube.com/watch?v=gWcaZs683Lk.

65 *Cleanerfishes curry favor*: Redouan Bshary and Manuela Würth, "Cleaner Fish *Labroides dimidiatus* Manipulate Client Reef Fish by Providing Tactile Stimulation," *Proceedings of the Royal Society of London B: Biological Sciences* 268 (2001): 1495–1501.

65 *diver Sean Payne described an encounter*: Jennifer S. Holland, *Unlikely*

Friendships: 47 Remarkable Stories from the Animal Kingdom (New York: Workman Publishing, 2011), 32.

66 *Andrea Marshall, the founder*: *Shark* [nature documentary series], BBC, 2015, www.bbc.co.uk/programmes/p02n7s0d.

66 *It's a similar story at the Shedd*: Karen Furnweger, "Shark Week: Sharks of a Different Stripe," Shedd Aquarium Blog, August 6, 2013, www.sheddaquarium.org/blog/2013/08/Shark-Week-Sharks-of-a-Different-Stripe.

67 *Dare we think the sunfish knows*: Tierney Thys, "Swimming Heads," *Natural History* 103 (1994): 36–39.

PART III: WHAT A FISH FEELS
PAIN, CONSCIOUSNESS, AND AWARENESS

69 *"Your life a sluice"*: D. H. Lawrence, "Fish."

71 *"Water wetly on fire"*: D. H. Lawrence,"Fish."

71 *limited opinion research on this question*: Caleb T. Hasler et al., "Opinions of Fisheries Researchers, Managers, and Anglers Towards Recreational Fishing Issues: An Exploratory Analysis for North America," *American Fisheries Symposium* 75 (2011): 141–70.

71 *and a survey of New Zealanders*: R. Muir et al., "Attitudes Towards Catch-and-Release Recreational Angling, Angling Practices and Perceptions of Pain and Welfare in Fish in New Zealand," *Animal Welfare* 22 (2013): 323–29.

74 *James Rose, a professor*: James D. Rose et al., "Can Fish Really Feel Pain?" *Fish and Fisheries* 15, no. 1 (2014): 97–133, published online December 20, 2012, doi:10.1111/faf.12010. As this manuscript was going to press, an article by Australian neuroscientist Brian Key titled "Why Fish Do Not Feel Pain" was published in the journal *Animal Sentience*, which generated a slew of formal commentaries (mostly rebuttals) published in the same journal, http://animalstudiesrepository.org/animsent.

75 *So impressive are the conscious acts*: Erich D. Jarvis et al., "Avian Brains and a New Understanding of Vertebrate Brain Evolution," *Nature Reviews Neuroscience* 6 (2005): 151–59.

75 *The fishes' answer*: O. R. Salva, V. A. Sovrano, and G. Vallortigara, "What Can Fish Brains Tell Us About Visual Perception?" *Frontiers in Neural Circuits* 8 (2014): 119, doi:10.3389/fncir.2014.00119.

76 *"Stories abound of bass that are caught"*: Keith A. Jones, *Knowing Bass: The Scientific Approach to Catching More Fish* (Guilford, CT: Lyons Press, 2001), 244.

76 *Carps and pikes avoided bait*: J. J. Beukema, "Acquired Hook-Avoidance in the Pike *Esox lucius L.* Fished with Artificial and Natural Baits," *Journal of Fish Biology* 2, no. 2 (1970): 155–60; J. J. Beukema, "Angling Experiments

with Carp (*Cyprinus carpio* L.) II. Decreased Catchability Through One Trial Learning," *Netherlands Journal of Zoology* 19 (1970): 81–92.

76 *A series of tests on largemouth basses*: R. O. Anderson and M. L. Heman, "Angling as a Factor Influencing the Catchability of Largemouth Bass," *Transactions of the American Fisheries Society* 98 (1969): 317–20.

76 *"They need to eat"*: Bruce Friedrich, "Toward a New Fish Consciousness: An Interview with Dr. Culum Brown," June 23, 2014, www.thedodo.com/community/FarmSanctuary/toward-a-new-fish-consciousness-601529872.html.

77 *Their findings are summarized*: Victoria A. Braithwaite, *Do Fish Feel Pain?* (Oxford: Oxford University Press, 2010); Lynne U. Sneddon, "The Evidence for Pain in Fish: The Use of Morphine as an Analgesic," *Applied Animal Behaviour Science* 83, no. 2 (2003): 153–62.

78 *But the proportion skew may*: L. U. Sneddon, "Pain in Aquatic Animals." *Journal of Experimental Biology* 218 (2015): 967–76.

79 *The trouts' negative reactions to the insults*: L. U. Sneddon, V. A. Braithwaite, and Michael J. Gentle, "Do Fishes Have Nociceptors? Evidence for the Evolution of a Vertebrate Sensory System," *Proceedings of the Royal Society B: Biological Sciences* 270 (2003): 1115–21; reported in Braithwaite, *Do Fish Feel Pain?*

79 *In separate experiments being conducted*: Lilia S. Chervova and Dmitri N. Lapshin, "Pain Sensitivity of Fishes and Analgesia Induced by Opioid and Nonopioid Agents," *Proceedings of the Fourth International Iran and Russia Conference* (Moscow: Moscow State University, 2004).

80 *Knowing this, the researchers built*: Braithwaite, *Do Fish Feel Pain?*, 68.

81 *For example, paradise fishes responded*: Vilmos Csányi and Judit Gervai, "Behavior-Genetic Analysis of the Paradise Fish, *Macropodus opercularis*. II. Passive Avoidance Learning in Inbred Strains," *Behavior Genetics* 16, no. 5 (1986): 553–57.

81 *In a study using 132 zebrafishes*: Caio Maximino, "Modulation of Nociceptive-like Behavior in Zebrafish (*Danio rerio*) by Environmental Stressors," *Psychology and Neuroscience* 4, no. 1 (2011): 149–55.

81 *Lynne Sneddon used what I consider*: L. U. Sneddon, "Clinical Anesthesia and Analgesia in Fish," *Journal of Exotic Pet Medicine* 21, no. 1 (2012): 32–43; "Do Painful Sensations and Fear Exist in Fish?" In *Animal Suffering: From Science to Law: International Symposium*, ed. Thierry Auffret Van der Kemp and Martine Lachance, 93–112 (Toronto: Carswell, 2013).

82 *When Janicke Nordgreen*: Janicke Nordgreen et al., "Thermonociception in Fish: Effects of Two Different Doses of Morphine on Thermal Threshold and Post-Test Behaviour in Goldfish (*Carassius auratus*)," *Applied Animal Behaviour Science* 119 (2009): 101–07.

83 *"Suggestions that finfish"*: *AVMA Guidelines for the Euthanasia of Animals: 2013 Edition*, American Veterinary Medical Association, www.avma.org/KB/Policies/Documents/euthanasia.pdf.

83 *In 2012 an august group of scientists*: Philip Low et al., "The Cambridge Declaration on Consciousness," proclaimed at the Francis Crick Memorial Conference on Consciousness in Human and Non-Human Animals, Cambridge, UK, July 7, 2012.

84 *psychologist and author Gay Bradshaw declared*: G. A. Bradshaw, "The Elephants Will Not Be Televised," *Psychology Today*, December 4, 2012, www.psychologytoday.com/blog/bear-in-mind/201212/the-elephants-will-not-be-televised.

84 *They can learn to avoid electric shocks*: Rudoph H. Ehrensing and Gary F. Michell, "Similar Antagonism of Morphine Analgesia by MIF-1 and Naloxone in *Carassius auratus*," *Pharmacology Biochemistry and Behavior* 17, no. 4 (1981): 757–61; Beukema, "Acquired Hook-Avoidance," "Angling Experiments with Carp."

FROM STRESS TO JOY

86 *"The fish's face is"*: Brian Curtis, *The Life Story of the Fish: His Manners and Morals* (New York: Harcourt Brace, 1949; repr. ed., Dover Publications, 1961).

87 *A separate account, published*: Joan Dunayer, *Animal Equality: Language and Liberation* (Derwood, MD: Ryce Publishing, 2001). Original source cited by Dunayer: Trevor Berry quoted in Robin Brown, "Blackie Was (Fin)ished until Big Red Swam In," *Weekend Argus* (Cape Town, South Africa), August 18, 1984: 15.

88 *Emotions involve relatively old brain circuits*: K. P. Chandroo, I. J. H. Duncan, and R. D. Moccia, "Can Fish Suffer? Perspectives on Sentience, Pain, Fear and Stress," *Applied Animal Behaviour Science* 86 (2004): 225–50; C. Broglio et al., "Hallmarks of a Common Forebrain Vertebrate Plan: Specialized Pallial Areas for Spatial, Temporal and Emotional Memory in Actinopterygian Fish," *Brain Research Bulletin* 66 (2005): 277–81; Eleanor Boyle, "Neuroscience and Animal Sentience," March 2009, www.ciwf.org.uk/includes/documents/cm_docs/2009/b/boyle_2009_neuroscience_and_animal_sentience.pdf.

88 *How the brain produces hormonal patterns*: F. A. Huntingford et al., "Current Issues in Fish Welfare," *Journal of Fish Biology* 68, no. 2 (2006): 332–72; S. E. Wendelaar Bonga, "The Stress Response in Fish," *Physiological Reviews* 77, no. 3 (1997): 591–625.

88 *Researchers from McMaster University*: Adam R. Reddon et al., "Effects of Isotocin on Social Responses in a Cooperatively Breeding Fish," *Animal Behaviour* 84 (2012): 753–60; "Swimming with Hormones: Researchers Unravel Ancient Urges That Drive the Social Decisions of Fish," McMaster University Press Release, October 9, 2012, www.eurekalert.org/pub_releases/2012-10/mu-swh100912.php.

89 *When this region is either disabled*: Chandroo et al., "Can Fish Suffer?"

89 *Studies on goldfishes have also shown*: Manuel Portavella, Blas Torres, and Cosme Salas, "Avoidance Response in Goldfish: Emotional and Temporal Involvement of Medial and Lateral Telencephalic Pallium," *Journal of Neuroscience* 24, no. 9 (2004): 2335–42.

89 *They respond as we might expect*: Chandroo et al., "Can Fish Suffer?"

89 *they also stop feeding*: Huntingford et al., "Current Issues in Fish Welfare."

90 *exposed them to oxazepam*: Jonatan Klaminder et al., "The Conceptual Imperfection of Aquatic Risk Assessment Tests: Highlighting the Need for Tests Designed to Detect Therapeutic Effects of Pharmaceutical Contaminants," *Environmental Research Letters* 9, no. 8 (2014): 084003.

90 *For example, naive fathead minnows*: D. P. Chivers and R. J. F. Smith, "Fathead Minnows, *Pimephales promelas,* Acquire Predator Recognition When Alarm Substance Is Associated with the Sight of Unfamiliar Fish," *Animal Behaviour* 48, no. 3 (1994): 597–605.

90 *Scientists from the University of Saskatchewan*: Adam L. Crane and Maud C. O. Ferrari, "Minnows Trust Conspecifics More Than Themselves When Faced with Conflicting Information About Predation Risk," *Animal Behaviour* 100 (2015): 184–90.

91 *from unsettling studies of rats, dogs*: Eighty published studies reviewed in J. P. Balcombe, Neal D. Barnard, and Chad Sandusky, "Laboratory Routines Cause Animal Stress," *Contemporary Topics in Laboratory Animal Science* 43, no. 6 (2004): 42–51.

91 *zebrafishes with a cortisol deficit*: L. Ziv et al., "An Affective Disorder in Zebrafish with Mutation of the Glucocorticoid Receptor," *Molecular Psychiatry* 18 (2013): 681–91.

92 *Do fishes seek ways to chill out?*: Chelsea Whyte, "Study: Fish Get a Fin Massage and Feel More Relaxed," *Washington Post,* November 21, 2011, www.washingtonpost.com/national/health-science/study-fish-get-a-fin-massage-and-feel-more-relaxed/2011/11/16/gIQAxoZvhN_story.html.

92 *Surmising that the caresses*: Marta C. Soares et al., "Tactile Stimulation Lowers Stress in Fish," *Nature Communications* 2 (2011): 534.

94 *the goldfish brain has cells containing dopamine*: Bow Tong Lett and Virginia L. Grant, "The Hedonic Effects of Amphetamine and Pentobarbital in Goldfish," *Pharmacological Biochemistry and Behavior* 32, no. 1 (1989): 355–56.

94 *Scientists have been exploring animal play*: Karl Groos, *The Play of Animals* (New York: Appleton and Company, 1898).

95 *the most comprehensive exploration*: Gordon M. Burghardt, *The Genesis of Animal Play: Testing the Limits* (Cambridge, MA: The MIT Press, 2005).

95 *interacting with this thermometer*: G. M. Burghardt, Vladimir Dinets, and James B. Murphy, "Highly Repetitive Object Play in a Cichlid Fish (*Tropheus duboisi*)," *Ethology* 121, no. 1 (2014): 38–44.

99 *The so-called* aerial respiration hypothesis: H. Dickson Hoese, "Jumping

Mullet—The Internal Diving Bell Hypothesis," *Environmental Biology of Fishes* 13, no. 4 (1985): 309–14.

99 *Gordon Burghardt published accounts*: Burghardt, *Genesis of Animal Play.*

PART IV: WHAT A FISH THINKS
FINS, SCALES, AND INTELLIGENCE

103 *"Nothing is too wonderful"*: Michael Faraday, laboratory journal entry #10,040 (19 March 1849), published in *The Life and Letters of Faraday Vol. II*, edited by Henry Bence Jones (Longmans, Green and Company, 1870), 253.

105 *"Every other animal currently considered stupid"*: Vladimir Dinets, *Dragon Songs: Love and Adventure among Crocodiles, Alligators and Other Dinosaur Relations* (New York: Arcade Publishing, 2013), 317.

107 *The use of cognitive maps*: Edward C. Tolman, "Cognitive Maps in Rats and Men," *The Psychological Review* 55, no. 4 (1948): 189–208.

107 *The goby's skill was demonstrated*: Lester R. Aronson, "Further Studies on Orientation and Jumping Behaviour in the Gobiid Fish, *Bathygobius soporator*," *Annals of the New York Academy of Sciences* 188 (1971): 378–92.

107 *brains of rock pool–dwelling goby*: G. E. White and C. Brown, "Microhabitat Use Affects Brain Size and Structure in Intertidal Gobies," *Brain, Behavior and Evolution* 85, no. 2 (2015): 107–16.

107 *As the biologist and author Vladimir Dinets*: V. Dinets, post on *r/science*, the forum of the *New Reddit Journal of Science*, November 6, 2014, www.reddit.com/r/science/comments/2lgxl6.

108 *Tony Pitcher, a biology professor*: Tony J. Pitcher, Foreword, *Fish Cognition and Behaviour*, ed. Culum Brown, Kevin Laland, and Jens Krause (Oxford: Wiley-Blackwell, 2006).

109 *In 1908, Jacob Reighard*: Jacob Reighard, "An Experimental Field-study of Warning Coloration in Coral Reef Fishes," *Papers from the Tortugas Laboratory of the Carnegie Institution of Washington*, vol. II (Washington, D.C.: Carnegie Institution, 1908): 257–325.

109 *adult crimson-spotted rainbowfishes*: Culum Brown, "Familiarity with the Test Environment Improves Escape Responses in the Crimson Spotted Rainbowfish, *Melanotaenia duboulayi*," *Animal Cognition* 4 (2001): 109–13.

110 *hook shyness by carps*: Beukema, "Acquired Hook-Avoidance," "Angling Experiments with Carp."

110 *paradise fishes who for several months avoided*: Vilmos Csányi and Antal Dóka, "Learning Interactions between Prey and Predator Fish," *Marine Behaviour and Physiology* 23 (1993): 63–78.

110 *When his usual dinner gong was reintroduced*: Zoe Catchpole, "Fish with a Memory for Meals Like a Pavlov Dog," *The Telegraph*, February 2, 2008, www.telegraph.co.uk/news/earth/earthnews/3323994/Fish-with-a-memory-for-meals-like-a-Pavlov-dog.html.

110 *"For almost every feat of learning"*: Reebs, *Fish Behavior*, 74.
111 *If you want to impress someone*: Chandroo et al., "Can Fish Suffer?"
111 *Culum Brown and his colleagues*: Culum Brown, unpublished data; Stéphan G. Reebs, "Time-Place Learning in Golden Shiners (Pisces: Cyprinidae)," *Behavioral Processes* 36, no. 3 (1996): 253–62.
112 *Within about two weeks*: Reebs, "Time-Place Learning"; L. M. Gómez-Laplaza and R. Gerlai, "Quantification Abilities in Angelfish (*Pterophyllum scalare*): The Influence of Continuous Variables," *Animal Cognition* 16 (2013): 373–83.
112 *By comparision, rats take slightly less*: Larry W. Means, S. R. Ginn, M. P. Arolfo, J. D. Pence, "Breakfast in the Nook and Dinner in the Dining Room: Time-of-day Discrimination in Rats," *Behavioral Processes*, 2000, 49: 21–33.
112 *garden warblers learn slightly more complex tasks*: Herbert Biebach, Marijke Gordijn, and John R. Krebs, "Time-and-Place Learning by Garden Warblers, *Sylvia borin*," *Animal Behaviour* 37, part 3 (1989): 353–60.
112 *Lacking the worldly survival skills*: W. J. McNeil, "Expansion of Cultured Pacific Salmon into Marine Ecosystems," *Aquaculture* 98 (1991): 123–30; www.usbr.gov/uc/rm/amp/twg/mtgs/03jun30/Attach_02.pdf.
112 *animals bred and reared in captivity*: Andrea S. Griffin, Daniel T. Blumstein, and Christopher S. Evans, "Training Captive-Bred or Translocated Animals to Avoid Predators," *Conservation Biology* 14 (2000): 1317–26.
112 *But when the biologists*: Flávia de Oliveira Mesquita and Robert John Young, "The Behavioural Responses of Nile Tilapia (*Oreochromis niloticus*) to Anti-Predator Training," *Applied Animal Behaviour Science* 106 (2007): 144–54.
113 *As early as the 1960s*: Lester R. Aronson, Frederick R. Aronson, and Eugenie Clark, "Instrumental Conditioning and Light-Dark Discrimination in Young Nurse Sharks," *Bulletin of Marine Science* 17, no. 2 (1967): 249–56.
113 *Demian Chapman with the Institute*: *Shark*, BBC, 2015, www.bbc.co.uk/programmes/p02n7s0d.
113 *problem solving by a cartilaginous fish*: Michael J. Kuba, Ruth A. Byrne, and Gordon M. Burghardt, "A New Method for Studying Problem Solving and Tool Use in Stingrays (*Potamotrygon castexi*)," *Animal Cognition* 13, no. 3 (2010): 507–13.
114 *And they show tool use*: Benjamin B. Beck, *Animal Tool Behavior: The Use and Manufacture of Tools by Animals* (New York: Taylor and Francis, 1980).
115 *how ecological challenges can affect intelligence*: K. K. Sheenaja and K. John Thomas, "Influence of Habitat Complexity on Route Learning Among Different Populations of Climbing Perch (*Anabas testudineus* Bloch, 1792)," *Marine and Freshwater Behaviour and Physiology* 44, no. 6 (2011): 349–58.
116 *"Even their previously bulging eyes"*: Lisa Davis, personal communication, September 2013.

116 *Using positive reinforcement to train fishes*: www.youtube.com/watch?v=Mbz1CaiqlYs shark Shedd Aquarium; www.youtube.com/watch?v=5k1FTrs0vno manta ray stretcher training.

TOOLS, PLANS, AND MONKEY MINDS

118 *"Knowledge comes, but wisdom lingers"*: Alfred, Lord Tennyson, "Locksley Hall," 1835.

118 *Bernardi's video unveils new gems*: Giacomo Bernardi, "The Use of Tools by Wrasses (Labridae)," *Coral Reefs* 31, no. 1 (2012): 39.

119 *The fish carried one of the pellets*: Łukasz Paśko, "Tool-like Behavior in the Sixbar Wrasse, *Thalassoma hardwicke* (Bennett, 1830)," *Zoo Biology* 29, no. 6 (2010): 767–73.

120 *Their eyes are sufficiently wide*: Robert W. Shumaker, Kristina R. Walkup, and Benjamin B. Beck, *Animal Tool Behavior: The Use and Manufacture of Tools by Animals*, rev. and updated ed. (Baltimore: Johns Hopkins University Press, 2011).

121 *But after watching a thousand attempts*: Stefan Schuster et al., "Animal Cognition: How Archer Fish Learn to Down Rapidly Moving Targets," *Current Biology* 16, no. 4 (2006): 378–83.

122 *Having a generalizable rule of fin*: Stefan Schuster et al., "Archer Fish Learn to Compensate for Complex Optical Distortions to Determine the Absolute Size of Their Aerial Prey," *Current Biology* 14, no. 17 (2004): 1565–68, doi:10.1016/j.cub.2004.08.050.

122 *Each fish wore a colored plastic tag*: Sandie Millot et al., "Innovative Behaviour in Fish: Atlantic Cod Can Learn to Use an External Tag to Manipulate a Self-Feeder," *Animal Cognition* 17, no. 3 (2014): 779–85.

123 *In January 2014 at Schroda Dam*: Gordon C. O'Brien et al., "First Observation of African Tigerfish *Hydrocynus vittatus* Predating on Barn Swallows *Hirundo rustica* in Flight," *Journal of Fish Biology* 84, no. 1 (2014): 263–66, doi:10.1111/jfb.12278.

126 *spending considerably more time foraging*: G. C. O'Brien et al., "A Comparative Behavioural Assessment of an Established and New Tigerfish (*Hydrocynus vittatus*) Population in Two Artificial Impoundments in the Limpopo Catchment, Southern Africa," *African Journal of Aquatic Sciences* 37, no. 3 (2012): 253–63.

126 *Introduced in 1983, they have survived*: Flora Malein, "Catfish Hunt Pigeons in France," Tech Guru Daily, December 10, 2012, www.tgdaily.com/general-sciences-features/67959-catfish-hunt-pigeons-in-france.

128 *Who do you think did better?*: Lucie H. Salwiczek et al., "Adult Cleaner Wrasse Outperform Capuchin Monkeys, Chimpanzees and Orangutans in a Complex Foraging Task Derived from Cleaner–Client Reef Fish Cooperation," *PLoS ONE* 7 (2012): e49068. doi:10.1371/journal.pone.0049068.

128 *tried the test on his four-year-old daughter*: Alison Abbott, "Animal Behaviour: Inside the Cunning, Caring and Greedy Minds of Fish," *Nature News*, May 26, 2015.

128 *The authors draw a key*: Salwiczek et al., "Adult Cleaner Wrasses Outperform Capuchin Monkeys," 3.

129 *chimpanzees far outperform humans*: Sana Inoue and Tetsuro Matsuzawa, "Working Memory of Numerals in Chimpanzees," *Current Biology* 17, no. 23 (2007): R1004–R1005.

129 *They also have the wits to*: You can watch a chimpanzee spontaneously use Archimedes' principle to solve a food puzzle on a video titled "Insight Learning: Chimpanzee Problem Solving" at: www.youtube.com/watch?v=fPz6uvIbWZE.

130 *Orangutans make mental maps of the locations*: Eugene Linden, *The Octopus and the Orangutan: Tales of Animal Intrigue, Intelligence and Ingenuity* (London: Plume, 2003).

130 *the concept of multiple intelligences*: Howard Gardner, *Frames of Mind: The Theory of Multiple Intelligences* (New York: Basic Books, 1983).

PART V: WHO A FISH KNOWS SUSPENDED TOGETHER

133 *"We of alien looks"*: C. J. Sansom, *Revelation: A Matthew Shardlake Tudor Mystery* (New York: Viking, 2009), 57.

135 *slime shed from the bodies*: McFarland, *Oxford Companion to Animal Behavior.*

135 *A subsequent study of wild-caught*: J. K. Parrish and W. K. Kroen, "Sloughed Mucus and Drag Reduction in a School of Atlantic Silversides, *Menidia menidia*," *Marine Biology* 97 (1988): 165–69.

136 *Familiar shoals of fathead minnows*: D. P. Chivers, G. E. Brown, and R. J. F. Smith, "Familiarity and Shoal Cohesion in Fathead Minnows (*Pimephales promelas*): Implications for Antipredator Behavior," *Canadian Journal of Zoology* 73, no. 5 (1995): 955–60.

136 *But when Krause introduced the fish alarm substance*: Jens Krause, "The Influence of Food Competition and Predation Risk on Size-assortative Shoaling in Juvenile Chub (*Leuciscus cephalus*)," *Ethology* 96, no. 2 (1994): 105–16.

136 *Position in the shoal*: McFarland, *Oxford Companion to Animal Behavior.*

137 *No wonder black or white mollies*: Scott P. McRobert and Joshua Bradner, "The Influence of Body Coloration on Shoaling Preferences in Fish," *Animal Behaviour* 56 (1998): 611–15.

137 *Avoiding conspicuousness may be another reason*: Jens Krause and Jean-Guy J. Godin, "Influence of Parasitism on Shoal Choice in the Banded Killifish (*Fundulus diaphanus*, Teleostei: Cyprinodontidae)," *Ethology* 102, no. 1 (1996): 40–49.

137 *Despite the speed of this movement*: e.g., McFarland, *Oxford Companion to Animal Behavior*.

137 *That probably explains why killifishes*: D. J. Hoare et al., "Context-Dependent Group Size Choice in Fish," *Animal Behaviour* 67, no. 1 (2004): 155–64.

138 *Not only can they*: Redouan Bshary, "Machiavellian Intelligence in Fishes," in *Fish Cognition and Behaviour*, C. Brown, K. Laland, and J. Krause, eds. (Oxford: Wiley-Blackwell, 2006).

138 *In captivity, European minnows*: McFarland, *Oxford Companion to Animal Behavior*.

138 *A smart guppy can know when*: Joseph Stromberg, "Are Fish Far More Intelligent Than We Realize?" Last updated August 4, 2014, www.vox.com/2014/8/4/5958871/fish-intelligence-smart-research-behavior-pain.

138 *guppies may use this knowledge*: Stromberg, "Are Fish Far More Intelligent . . . ?"

139 *the east African freshwater cichlid*: Logan Grosenick, Tricia S. Clement, and Russell D. Fernald, "Fish Can Infer Social Rank by Observation Alone," *Nature* 445 (2007): 429–32.

139 *Individual fishes plucked from groups*: Neil B. Metcalfe and Bruce C. Thomson, "Fish Recognize and Prefer to Shoal with Poor Competitors," *Proceedings of the Royal Society of London B: Biological Sciences* 259 (1995): 207–10.

139 *Bluegill sunfishes, and probably many other*: Lee Alan Dugatkin and D. S. Wilson, "The Prerequisites for Strategic Behavior in Bluegill Sunfish, *Lepomis macrochirus*," *Animal Behaviour* 44 (1992): 223–30.

140 *"Absolutely. I'm the one"*: Pete Brockdor, personal communication, April 12, 2014.

140 *When presented with two human faces*: C. Newport, G. M. Wallis, and U. E. Siebeck, "Human Facial Recognition in Fish," European Conference on Visual Perception (ECVP) Abstracts, *Perception* 42, no. 1 suppl (2013): 160.

141 *Territoriality is widespread among fishes*: Helfman and Collette, *Fishes: The Animal Answer Guide*.

141 *Remarkably, Godard discovered*: Renee Godard, "Long-Term Memory of Individual Neighbours in a Migratory Songbird," *Nature* 350 (1991): 228–29.

142 *a simple and effective method*: Ronald E. Thresher, "The Role of Individual Recognition in the Territorial Behaviour of the Threespot Damselfish, *Eupomacentrus planifrons*," *Marine Behaviour and Physiology* 6, no. 2 (1979): 83–93.

143 *During territorial disputes, a pair*: Roldan C. Muñoz et al., "Extraordinary Aggressive Behavior from the Giant Coral Reef Fish, *Bolbometopon muricatum*, in a Remote Marine Reserve," *PLoS ONE* 7, no. 6 (2012): e38120, doi:10.1371/journal.pone.0038120.

143 *As bumpheads become scarcer*: Muñoz et al., "Extraordinary Aggressive Behavior."

SOCIAL CONTRACTS

151 *"One hand washes the other"*: Seneca ("*Manus manum lavet*").

152 *Freshwater cleanerfishes include cichlids*: Alexandra S. Grutter, "Cleaner Fish," *Current Biology* 20, no. 13 (2010): R547–R549.

152 *Other clients include lobsters*: Grutter, "Cleaner Fish"; McFarland, *Oxford Companion to Animal Behavior.*

152 *A study on the Great Barrier Reef*: A. S. Grutter, "Parasite Removal Rates by the Cleaner Wrasse *Labroides dimidiatus*," *Marine Ecology Progress Series* 130 (1996): 61–70.

152 *Some individual clients visited*: A. S. Grutter, "The Relationship between Cleaning Rates and Ectoparasite Loads in Coral Reef Fishes," *Marine Ecology Progress Series* 118 (1995): 51–58.

153 *When Grutter thwarted access to cleanerfishes*: A. S. Grutter, Jan Maree Murphy, and J. Howard Choat, "Cleaner Fish Drives Local Fish Diversity on Coral Reefs," *Current Biology* 13, no. 1 (2003): 64–67.

153 *This sort of species decline*: A. S. Grutter, "Effect of the Removal of Cleaner Fish on the Abundance and Species Composition of Reef Fish," *Oecologia* 111, no. 1 (1997): 137–43.

153 *Cleaners also vibrate their ventral fins*: McFarland, *Oxford Companion to Animal Behavior.*

154 *If a cleaner is in the gills*: Desmond Morris, *Animalwatching: A Field Guide to Animal Behavior* (New York: Crown Publishers, 1990).

154 *The cleaners show no fear*: *Shark* [documentary series], BBC, www.bbc.co.uk/programmes/p02n7s0d.

154 *With dozens of clients*: Sabine Tebbich, Redouan Bshary, and Alexandra S. Grutter, "Cleaner Fish *Labroides dimidiatus* Recognise Familiar Clients," *Animal Cognition* 5, no. 3 (2002): 139–45.

154 *clients showed no such preference*: Tebbich et al., "Cleaner Fish *Labroides dimidiatus* Recognise . . ."

154 *It reminds me of hummingbirds' ability*: Melissa Bateson, Susan D. Healy, and T. Andrew Hurly, "Context-Dependent Foraging Decisions in Rufous Hummingbirds," *Proceedings of the Royal Society of London B: Biological Sciences* 270 (2003): 1271–76. www.jstor.org/stable/3558811?seq=1#page_scan_tab_contents.

155 *By using memory along three dimensions*: Lucie H. Salwiczek and Redouan Bshary, "Cleaner Wrasses Keep Track of the 'When' and 'What' in a Foraging Task," *Ethology* 117, no. 11 (2011): 939–48.

155 *According to a study*: Jennifer Oates, Andrea Manica, and Redouan Bshary, "The Shadow of the Future Affects Cooperation in a Cleaner Fish," *Current Biology* 20, no. 11 (2010): R472–R473.

156 *They do this by facing away*: Bshary and Würth, "Cleaner Fish *Labroides dimidiatus* Manipulate."

156 *Cleaners are more likely to caress*: Bshary and Würth.
156 *are considered safe havens*: Karen L. Cheney, R. Bshary, A. S. Grutter, "Cleaner Fish Cause Predators to Reduce Aggression Towards Bystanders at Cleaning Stations," *Behavioural Ecology* 19, no. 5 (2008): 1063–67.
156 *By doing this, client fishes*: Bshary, "Machiavellian Intelligence in Fishes."
157 *No wonder cleaners behave more cooperatively*: R. Bshary, Arun D'Souza, "Cooperation in Communication Networks: Indirect Reciprocity in Interactions Between Cleaner Fish and Client Reef Fish," in *Animal Communication Networks*, ed. Peter K. McGregor, 521–39 (Cambridge: Cambridge University Press, 2005).
157 *But a resident client*: R. Bshary, A. S. Grutter, "Asymmetric Cheating Opportunities and Partner Control in the Cleaner Fish Mutualism," *Animal Behaviour* 63, no. 3 (2002): 547–55.
157 *Punishment has been shown*: Bshary, "Machiavellian Intelligence in Fishes."
157 *This shift in cleaner behavior*: Marta C. Soares et al., "Does Competition for Clients Increase Service Quality in Cleaning Gobies?" *Ethology* 114, no. 6 (2008): 625–32.
159 *It turns out that blennies*: Andrea Bshary and Redouan Bshary, "Self-Serving Punishment of a Common Enemy Creates a Public Good in Reef Fishes," *Current Biology* 20, no. 22 (2010): 2032–35.
160 *My speakers attracted crowds*: J. P. Balcombe and M. Brock Fenton, "Eavesdropping by Bats: The Influence of Echolocation Call Design and Foraging Strategy," *Ethology* 79, no. 2 (1988): 158–66.
160 *Warner set about removing*: Robert R. Warner, "Traditionality of Mating-Site Preferences in a Coral Reef Fish," *Nature* 335 (1988): 719–21, 719.
161 *Others include herrings, groupers*: Helfman et al., *Diversity of Fishes* (2009).
161 *Their chosen path even persists*: Culum Brown and Kevin M. Laland, "Social Learning in Fishes," in *Fish Cognition and Behaviour*, 186–202.
162 *They found that social cohesion*: Giancarlo De Luca et al., "Fishing Out Collective Memory of Migratory Schools," *Journal of the Royal Society Interface* 11, no. 95 (2014), doi:10.1098/rsif.2014.0043.
162 *North Atlantic right whales*: International Whaling Commission (undated), "Status of Whales," accessed November 29, 2014, http://iwc.int/status.
162 *As nets and hooks were turned*: www.terranature.org/orange_roughy.htm; www.eurekalert.org/pub_releases/2007-02/osu-ldf021307.php.

COOPERATION, DEMOCRACY, AND PEACEKEEPING

164 *"Nothing truly valuable"*: Albert Einstein, *The World As I See It* (Minneapolis, MN: Filiquarian Publishing, 2005), 44.
165 *barracuda will swim in a tight spiral*: Brian L. Partridge, Jonas Johansson, and John Kalish, "The Structure of Schools of Giant Bluefin Tuna in Cape Cod Bay," *Environmental Biology of Fishes* 9 (1983): 253–62.

166 *Cooperators had higher success rates*: Oona M. Lönnstedt, Maud C. O. Ferrari, and Douglas P. Chivers, "Lionfish Predators Use Flared Fin Displays to Initiate Cooperative Hunting," *Biology Letters* 10, no. 6 (2014), doi:10.1098/rsbl.2014.0281.

166 *Chasers flush the prey*: Carine Strübin, Marc Steinegger, and R. Bshary, "On Group Living and Collaborative Hunting in the Yellow Saddle Goatfish (*Parupeneus cyclostomus*)," *Ethology* 117, no. 11 (2011), 961–69.

166 *The reason for this success*: R. Bshary et al., "Interspecific Communicative and Coordinated Hunting Between Groupers and Giant Moray Eels in the Red Sea," *PLoS Biology* 4 (2006): e431.

167 *Commenting on the collaboration*: Frans B. M. de Waal, "Fishy Cooperation," *PLoS Biology* 4 (2006): e444, doi:10.1371/journal.pbio.0040444.

167 *The headstand signal meets the five criteria*: Alexander L. Vail, Andrea Manica, and R. Bshary, "Referential Gestures in Fish Collaborative Hunting," *Nature Communications* 4 (2013): 1765, doi:10.1038/ncomms2781; Simone Pika and Thomas Bugnyar, "The Use of Referential Gestures in Ravens (*Corvus corax*) in the Wild," *Nature Communications* 2 (2011): 560.

168 *But by the second day*: A. L. Vail, A. Manica, and R. Bshary, "Fish Choose Appropriately When and with Whom to Collaborate," *Current Biology* 24, no. 17 (2014): R791–R793, doi:10.1016/j.cub.2014.07.033.

169 *A grouper has no hands*: Ed Yong, "When Your Prey's in a Hole and You Don't Have a Pole, Use a Moray," http://phenomena.nationalgeographic .com/2014/09/08/when-your-preys-in-a-hole-and-you-dont-have-a-pole -use-a-moray.

169 *And this highly democratic process*: Jon Hamilton, "In Animal Kingdom, Voting of a Different Sort Reigns," NPR Online, last updated October 25, 2012, www.npr.org/2012/10/24/163561729/in-animal-kingdom-voting-of -a-different-sort-reigns3.

170 *It appears that groups either aggregate*: Iain D. Couzin, "Collective Cognition in Animal Groups," *Trends in Cognitive Sciences* 13, no. 1 (2009): 36–43; Larissa Conradt and Timothy J. Roper, "Consensus Decision Making in Animals," *Trends in Ecology and Evolution* 20, no. 8 (2005): 449–56.

170 *The sticklebacks behaved as if*: David J. T. Sumpter et al., "Consensus Decision Making by Fish," *Current Biology* 18 (2008): 1773–77.

170 *singleton sticklebacks are susceptible*: Ashley J. W. Ward et al., "Quorum Decision-Making Facilitates Information Transfer in Fish Shoals," *PNAS* 105, no. 19 (2008): 6948–53.

171 *Similarly, small schools of mosquitofishes*: A. J. W. Ward et al., "Fast and Accurate Decisions Through Collective Vigilance in Fish Shoals," *PNAS* 108, no. 6 (2011): 2312–15.

171 *actual physical fighting between rivals*: I discuss this at some length in Balcombe, *Second Nature: The Inner Lives of Animals* (New York: Palgrave Macmillan, 2010).

171 *fishes often use ritualized displays*: John Maynard-Smith and George Price, "The Logic of Animal Conflict," *Nature* 246 (1973): 15–18.
171 *Other embellishments include head shakes*: Reebs, *Fish Behavior.*
171 *The aggressively territorial blunthead cichlid*: McFarland, *Oxford Companion to Animal Behavior.*
172 *If one of the females is unfamiliar*: Mark H. J. Nelissen, "Structure of the Dominance Hierarchy and Dominance Determining 'Group Factors' in *Melanochromis auratus* (Pisces, Cichlidae)," *Behaviour* 94 (1985): 85–107.
172 *In an admirable show of restraint*: Marian Y. L. Wong et al., "The Threat of Punishment Enforces Peaceful Cooperation and Stabilizes Queues in a Coral-Reef Fish," *Proceedings of the Royal Society of London B: Biological Sciences* 274 (2007): 1093–99.
172 *dieting has been shown to improve*: M. Y. L. Wong et al., "Fasting or Feasting in a Fish Social Hierarchy," *Current Biology* 18, no. 9 (2008): R372–R373.
173 *Males who had watched*: Rui F. Oliveira, Peter K. McGregor, and Claire Latruffe, "Know Thine Enemy: Fighting Fish Gather Information from Observing Conspecific Interactions," *Proceedings of the Royal Society of London B: Biological Sciences* 265 (1998): 1045–49.
174 *Scientists reporting the association*: L. A. Rocha, R. Ross, and G. Kopp, "Opportunistic Mimicry by a Jawfish," *Coral Reefs* 31 (2011): 285, doi:10.1007/s00338-011-0855-y.
175 *When an inquisitive scavenger*: Ron Harlan, "Ten Devastatingly Deceptive or Bizarre Animal Mimics," Listverse, July 20, 2013, http://listverse.com/2013/07/20/10-devastatingly-deceptive-or-bizarre-animal-mimics.
175 *The small fishes they are hoping*: McFarland, *Oxford Companion to Animal Behavior.*
175 *Trumpetfishes will also join*: Morris, *Animalwatching.*
175 *in the perpetually dark oceanic abyss*: Pietsch, *Oceanic Anglerfishes.*

PART VI: HOW A FISH BREEDS
SEX LIVES

179 *"How do you spell 'love'?"*: A. A. Milne, *Winnie-the-Pooh* (New York: Puffin Books, 1992).
181 *"sexual plasticity and flexibility"*: T. J. Pandian, *Sexuality in Fishes* (Enfield, NH: Science Publishers, 2011).
181 *a Full Monty of breeding systems*: James S. Diana, *Biology and Ecology of Fishes, 2nd ed.* (Traverse City, MI: Biological Sciences Press/Cooper Publishing, 2004).
182 *there are scores of fishes*: Yvonne Sadovy de Mitcheson and Min Liu, "Functional Hermaphroditism in Teleosts," *Fish and Fisheries* 9, no. 1 (2008): 1–43.
182 *For example, in a mating system*: Robert R. Warner, "Mating Behavior and

Hermaphroditism in Coral Reef Fishes," *American Scientist* 72, no. 2 (1984): 128–36.

182 *who studied this strict mating system*: Hans Fricke and Simone Fricke, "Monogamy and Sex Change by Aggressive Dominance in Coral Reef Fish," *Nature* 266 (1977): 830–32.

183 *reveals a slight inaccuracy*: Helfman et al., *Diversity of Fishes* (2009), 458.

183 *how this happens is not clear*: Arimune Munakata and Makito Kobayashi, "Endocrine Control of Sexual Behavior in Teleost Fish," *General and Comparative Endocrinology* 165, no. 3 (2010): 456–68.

184 *what might have created this exquisite curiosity*: Some of Yoji Ookata's photos of this phenomenon, posted September 23, 2012, can be found here: http://mostlyopenocean.blogspot.com.au/2012/09/a-little-fish-makes-big-sand-sculptures.html.

185 *Almost as soon as eggs are laid*: Helfman et al., *Diversity of Fishes* (2009).

185 *Humans and bowerbirds are not*: Sara Östlund-Nilsson and Mikael Holmlund, "The Artistic Three-Spined Stickleback (*Gasterosteus aculeatus*)," *Behavioral Ecology and Sociobiology* 53, no. 4 (2003): 214–20.

186 *A female, in contrast, may do better*: Lesley Evans Ogden, "Fish Faking Orgasms and Other Lies Animals Tell for Sex," BBC Earth, February 14, 2015, www.bbc.com/earth/story/20150214-fake-orgasms-and-other-sex-lies?ocid=fbert.

187 *described as an apparent visual deception*: Norman and Greenwood, *History of Fishes*.

187 *The sperm pass rapidly through*: Masanori Kohda et al., "Sperm Drinking by Female Catfishes: A Novel Mode of Insemination," *Environmental Biology of Fishes* 42, no. 1 (1995): 1–6.

187 *timed the sperm's passage*: Kohda et al.

188 *The mussel's eggs adhere*: Morris, *Animalwatching*.

189 *It is thought that a male molly does this*: Martin Plath et al., "Male Fish Deceive Competitors About Mating Preferences," *Current Biology* 18, no. 15 (2008): 1138–41.

190 *influenced by the preferences of rivals*: Ingo Schlupp and Michael J. Ryan, "Male Sailfin Mollies (*Poecilia latipinna*) Copy the Mate Choice of Other Males," *Behavioral Ecology* 8, no. 1 (1997): 104–07.

191 *the gonopodium is directed backward*: Norman and Greenwood, *History of Fishes*.

191 *In some species the priapium*: Lois E. TeWinkel, "The Internal Anatomy of Two Phallostethid Fishes," *Biological Bulletin* 76, no. 1 (1939): 59–69.

191 *Careful anatomical study confirms*: Ralph J. Bailey, "The Osteology and Relationships of the Phallostethid Fishes," *Journal of Morphology* 59, no. 3 (2005): 453–83.

192 *the male with the longer organ*: R. Brian Langerhans, Craig A. Layman, and Thomas J. DeWitt, "Male Genital Size Reflects a Tradeoff Between

Attracting Mates and Avoiding Predators in Two Live-Bearing Fish Species," *PNAS* 102, no. 21 (2005): 7618–23.

193 *This odyssey has a Romeo*: Norman and Greenwood, *History of Fishes*.

193 *Captive lemonpeel angelfishes*: Ike Olivotto et al., "Spawning, Early Development, and First Feeding in the Lemonpeel Angelfish *Centropyge flavissimus*," *Aquaculture* 253 (2006): 270–78.

PARENTING STYLES

194 *"No one is useless"*: Charles Dickens, *Our Mutual Friend* (Oxford: Oxford University Press, 1989).

195 *Despite what we're taught*: Norman and Greenwood, *History of Fishes*.

195 *at least some form of caregiving*: Clive Roots, *Animal Parents* (Westport, CT: Greenwood Press, 2007); Judith E. Mank, Daniel E. L. Promislow, and John C. Avise, "Phylogenetic Perspectives in the Evolution of Parental Care in Ray-Finned Fishes," *Evolution* 59, no. 7 (2005): 1570–78.

195 *Some sharks have a placenta*: William C. Hamlett, "Evolution and Morphogenesis of the Placenta in Sharks," *Journal of Experimental Zoology* 252, Supplement S2 (1989): 35–52.

195 *produce body substances that act as food*: Helfman et al., *Diversity of Fishes* (1997).

195 *During several weeks of caring*: Norman and Greenwood, *History of Fishes*.

195 *A new family of peptide antibiotics*: Edward J. Noga and Umaporn Silphaduang, "Piscidins: A Novel Family of Peptide Antibiotics from Fish," *Drug News and Perspectives* 16, no. 2 (2003): 87–92.

196 *Tierney Thys, a world expert*: Thys, "For the Love of Fishes."

196 *Finally, they remove any sand grains*: Thys, personal communication, August 2015.

197 *the female's pelvic fins fuse*: Eleanor Bell, "Gasterosteiform," *Encyclopedia Britannica*, www.britannica.com/animal/gasterosteiform.

197 *The parent rolls in the egg mass*: McFarland, *Oxford Companion to Animal Behavior*.

198 *It must be an exhausting job*: C. O'Neil Krekorian and D. W. Dunham, "Preliminary Observations on the Reproductive and Parental Behavior of the Spraying Characid *Copeina arnoldi* Regan," *Zeitschrift für Tierpsychologie* 31, no. 4 (1972): 419–37.

198 *Pricklebacks, gunnels, and wolf eels*: Lawrence S. Blumer, "A Bibliography and Categorization of Bony Fishes Exhibiting Parental Care," *Zoological Journal of the Linnean Society* 76 (1982): 1–22.

198 *There must be advantages: higher incubation temperatures*: Helfman et al., *Diversity of Fishes* (2009).

199 *occurs in at least nine fish families*: Clive Roots, *Animal Parents*.

200 *mouthbrooders have been known to starve*: Andrew S. Hoey, David R. Bellwood, and Adam Barnett, "To Feed or to Breed: Morphological Constraints of Mouthbrooding in Coral Reef Cardinalfishes," *Proceedings of the Royal Society of London B: Biological Sciences* 279 (2012): 2426–32.

200 *They take no food*: Yasunobu Yanagisawa and Mutsumi Nishida, "The Social and Mating System of the Maternal Mouthbrooder *Tropheus moorii* in Lake Tanganyika," *Japanese Journal of Ichthyology* 38, no. 3 (1991): 271–82.

200 *Sensing danger, it is he*: Reebs, *Fish Behavior*.

200 *the heads of nine cardinalfish species*: Hoey et al., "To Feed or to Breed."

201 *"As ocean temperatures warm"*: "Saving the World's Fisheries," unsigned editorial, *Washington Post*, October 3, 2012.

201 *The female releases her eggs*: Roots, *Animal Parents*.

201 *females are also playing a numbers game*: Adam G. Jones and John C. Avise, "Sexual Selection in Male-Pregnant Pipefishes and Seahorses: Insights from Microsatellite Studies of Maternity," *Journal of Heredity* 92, no. 2 (2001): 150–58.

202 *Cooperative breeding is known from several*: Julie K. Desjardins et al., "Sex and Status in a Cooperative Breeding Fish: Behavior and Androgens," *Behavioral Ecology and Sociobiology* 62, no. 5 (2007): 785–94.

202 *Helpers perform a variety of tasks*: Helfman et al., *Diversity of Fishes* (2009).

203 *For birds, supportive evidence*: Jan Komdeur, "Importance of Habitat Saturation and Territory Quality for Evolution of Cooperative Breeding in the Seychelles Warbler," *Nature* 358 (1992): 493–95.

203 *Swiss researchers from the University of Bern*: Ralph Bergmüller, Dik Heg, and Michael Taborsky, "Helpers in a Cooperatively Breeding Cichlid Stay and Pay or Disperse and Breed, Depending on Ecological Constraints," *Processes in Biological Science* 272 (2005): 325–31.

205 *In over a quarter of clutches*: Rick Bruintjes et al., "Paternity of Subordinates Raises Cooperative Effort in Cichlids," *PLoS ONE* 6, no. 10 (2011): e25673, doi:10.1371/journal.pone.0025673.

205 *Genetic data collected from groups*: K. A. Stiver et al., "Mixed Parentage in *Neolamprologus pulcher* Groups," *Journal of Fish Biology* 74, no. 5 (2009): 1129–35, doi:10.1111/j.1095-8649.2009.02173.x.

205 *they have a higher genetic stake*: Bruintjes et al., "Paternity of Subordinates."

205 *After being returned to the nest site*: Bergmüller et al., "Helpers in a Cooperatively Breeding Cichlid"; R. Bergmüller, M. Taborsky, "Experimental Manipulation of Helping in a Cooperative Breeder: Helpers 'Pay to Stay' by Pre-emptive Appeasement," *Animal Behaviour* 69, no. 1 (2005): 19–28.

206 *there is evidence for a beneficial trade-off*: Michael S. Webster, "Interspecific Brood Parasitism of Montezuma Oropendolas by Giant Cowbirds: Parasitism or Mutualism?" *Condor* 96 (1994); 794–98.

206 *The father collects invertebrates*: Jay R. Stauffer and W. T. Loftus, "Brood

Parasitism of a Bagrid Catfish (*Bagrus meridionalis*) by a Clariid Catfish (*Bathyclarias nyasensis*) in Lake Malaŵi, Africa," *Copeia* 2010, no. 1: 71–74.

207 *Adding insult to audacity*: Tetsu Sato, "A Brood Parasitic Catfish of Mouthbrooding Cichlid Fishes in Lake Tanganyika," *Nature* 323 (1986): 58–59.

PART VII: FISH OUT OF WATER

209 *"I, a many-fingered horror"*: D. H. Lawrence, "Fish."

211 *The earliest known fishing net*: Arto Miettinen et al., "The Palaeoenvironment of the Antrea Net Find," in *Karelian Isthmus: Stone Age Studies in 1998–2003*, ed. Mika Lavento and Kerkko Nordqvist, 71–87 (Helsinki: The Finnish Antiquarian Society, 2008).

212 *"Though every year fish are taken"*: H. J. Shepstone, "Fishes That Come to the Deep-Sea Nets," in *Animal Life of the World*, ed. J. R Crossland and J. M. Parrish (London: Odhams Press, 1934), 525.

212 *the average human in 2009 was consuming*: FAO, "State of World Fisheries, Aquaculture Report—Fish Consumption" (2012), www.thefishsite.com/articles/1447/fao-state-of-world-fisheries-aquaculture-report-fish-consumption.

212 *In the United States, per capita*: Carrie R. Daniel et al., "Trends in Meat Consumption in the United States," *Public Health Nutrition* 14, no. 4 (2011): 575–83.

212 *Global fish numbers are shrinking*: Gaia Vince, "How the World's Oceans Could Be Running Out of Fish," September 21, 2012, www.bbc.com/future/story/20120920-are-we-running-out-of-fish.

212 *"Anybody who thinks there can be limitless growth"*: Adam Sherwin, "'Leave the badgers alone,' says Sir David Attenborough. 'The real problem is the human population,'" *The Independent*, November 5, 2012, www.independent.co.uk/environment/nature/leave-the-badgers-alone-says-sir-david-attenborough-the-real-problem-is-the-human-population-8282959.html.

213 *In longline fishing, lines with 2,500*: J. Rice, J. Cooper, P. Medley, and A. Hough, "South Georgia Patagonian Toothfish Longline Fishery," Moody Marine Ltd. (2006), www.msc.org/track-a-fishery/fisheries-in-the-program/certified/south-atlantic-indian-ocean/south-georgia-patagonian-toothfish-longline/assessment-documents/document-upload/SurvRep2.pdf.

213 *Fishes of all ages and sizes*: W. Jeffrey Bolster, *The Mortal Sea: Fishing the Atlantic in the Age of Sail* (Cambridge, MA: Belknap Press/Harvard University Press, 2012).

213 *The celebrated American oceanographer*: Lloyd Evans, "Making Waves: An Audience with Sylvia Earle, the Campaigner Known as Her Deepness," *The Spectator*, June 25, 2011, http://new.spectator.co.uk/2011/06/making-waves-2.

213 *They stay at sea for weeks*: FAO, "The Tuna Fishing Vessels of the World,"

chapter 4 of the FAO's "Managing Fishing Capacity of the World Tuna Fleet" (2003), www.fao.org/docrep/005/y4499e/y4499e07.htm.

213 *And there are lots of these factory ships*: FAO Fisheries Circular No. 949 FIIT/C949, "Analysis of the Vessels Over 100 Tons in the Global Fishing Fleet" (1999), www.fao.org/fishery/topic/1616/en.

214 *Fish-farming (a subset of aquaculture)*: J. Lucas, "Aquaculture," *Current Biology* 25 (2015): R1-R3; Lucas, personal communication, January 6, 2016.

214 *On trout farms, densities*: Philip Lymbery, "In Too Deep—Why Fish Farming Needs Urgent Welfare Reform" (2002), www.ciwf.org.uk/includes/documents/cm_docs/2008/i/in_too_deep_summary_2001.pdf.

214 *More than half the world's*: FAO, "Highlights of Special Studies," *The State of World Fisheries and Aquaculture 2008* (Rome: FAO, 2008), ftp://ftp.fao.org/docrep/fao/011/i0250e/i0250e03.pdf.

215 *According to one analysis*: Rosamond L. Naylor et al., "Effect of Aquaculture on World Fish Supplies," *Nature* 405 (2000): 1017–24.

215 *A catch cap on Atlantic menhadens*: P. Baker, "Atlantic Menhaden Catch Cap a Success," The Pew Charitable Trusts, May 15, 2014, www.pewtrusts.org/en/research-and-analysis/analysis/2014/05/15/atlantic-menhaden-catch-cap-a-success-millions-more-of-the-most-important-fish-in-the-sea.

215 *Most of the menhaden meal*: Jacqueline Alder et al., "Forage Fish: From Ecosystems to Markets," *Annual Review of Environment and Resources* 33 (2008): 153–66; Sylvester Hooke, "Fished Out! Scientists Warn of Collapse of all Fished Species by 2050," *Healing Our World* (Hippocrates Health Institute magazine) 32, no. 3 (2012): 28–29, 63.

215 *One company, Omega Protein*: Helfman and Collette, *Fishes: The Animal Answer Guide*.

216 *Overall death rates of 10 percent to*: Lymbery, "In Too Deep" (2002); www.ciwf.org.uk/includes/documents/cm_docs/2008/i/in_too_deep_summary_2001.pdf.

216 *impact salmon-dependent wildlife*: Cornelia Dean, "Saving Wild Salmon, in Hopes of Saving the Orca," *New York Times,* November 4, 2008.

216 *A single farm in Lake Nicaragua*: Elisabeth Rosenthal, "Another Side of Tilapia, the Perfect Factory Fish," *New York Times,* May 2, 2011.

217 *the young fishes have had no chance*: Culum Brown, T. Davidson, and K. Laland, "Environmental Enrichment and Prior Experience of Live Prey Improve Foraging Behavior in Hatchery-Reared Atlantic Salmon," *Journal of Fish Biology* 63, supplement S1 (2003):187–96.

217 *fishes' observational learning*: see Culum Brown, "Fish Intelligence, Sentience, and Ethics," *Animal Cognition,* (2014) 18:1–17.

219 *half a million fishes if they are herrings*: Based on average herring weights, and that a single set may contain 200 tons of herring. See, the Gulf of Maine Research Institute, www.gma.org/herring/harvest_and_processing/seining/default.asp.

219 *esophageal eversion*: Emily S. Munday, Brian N. Tissot, Jerry R. Heidel, and Tim Miller-Morgan, "The Effects of Venting and Decompression on Yellow Tang (*Zebrasoma flavescens*) in the Marine Ornamental Aquarium Fish Trade," *PeerJ* 3: e756, DOI 10.7717/peerj.756.

220 *banned in Germany as inhumane for killing eels*: Anon. (1997). Verordnung zum Schutz von Tieren in Zusammenhang mit der Schlachtung oder Tötung—TierSchlV (Tierschutz-Schlachtverordnung), vom 3. März 1997, Bundesgesetzblatt Jahrgang 1997 Teil I S. 405, zuletzt geändert am 13. April 2008 durch Bundesgesetzblatt Jahrgang 2008 Teil I Nr. 18, S. 855, Art. 19 vom 24. April 2006.

220 *Some of these methods*: D. H. F. Robb and S. C. Kestin, "Methods Used to Kill Fish: Field Observations and Literature Reviewed," *Animal Welfare* 11, no. 3 (2002): 269–82.

221 *That's the* daily *bycatch*: R. W. D. Davies et al., "Defining and Estimating Global Marine Fisheries Bycatch," *Marine Policy* 33, no. 4 (2009): 661–72.

221 *global yearly bycatch rates*: FAO Fisheries and Aquaculture Department, "Reduction of Bycatch and Discards," www.fao.org/fishery/topic/14832/en, accessed September 9, 2015.

221 *By this definition, bycatch today*: Davies et al., "Defining and Estimating Global Marine Fisheries Bycatch."

221 *Unwanted-fish-to-shrimp weight*: Helfman et al., *Diversity of Fishes* (2009).

221 *Overall, 105 fish species*: Helfman et al. (2009).

221 *Fishing fleets abandon or lose*: A. Butterworth, I. Clegg, and C. Bass, *Untangled—Marine Debris: A Global Picture of the Impact on Animal Welfare and of Animal-Focused Solutions* (London: World Society for the Protection of Animals [now: World Animal Protection], 2012).

222 *brought the dolphin kill rate down*: NOAA Fisheries, "The Tuna-Dolphin Issue," last modified December 24, 2014, https://swfsc.noaa.gov/textblock.aspx?Division=PRD&ParentMenuId=228&id=1408.

222 *But dolphin populations have not recovered*: Paul R. Wade et al., "Depletion of Spotted and Spinner Dolphins in the Eastern Tropical Pacific: Modeling Hypotheses for Their Lack of Recovery," *Marine Ecology Progress Series* 343 (2007), 1–14.

222 *Baited longlines and wire struts*: "Rosy Outlook," *New Scientist*, February 28, 2009, p 5.

222 *such simple bird-scaring designs are now recommended*: Agreement on the Conservation of Albatrosses and Petrels, "Best Practice Seabird Bycatch Mitigation," September 19, 2014, http://acap.aq/en/bycatch-mitigation/mitigation-advice/2595-acap-best-practice-seabird-bycatch-mitigation-criteria-and-definition/file.

224 *A shark's fins are not the only source*: [Wilcox 2015]. Christie Wilcox, "Shark fin ban masks growing appetite for its meat," www.theguardian.com/environment/2015/sep/12/shark-fin-ban-not-saving-species.

224 *Small wonder that some shark*: Juliet Eilperin, *Demon Fish: Travels Through the Hidden World of Sharks* (New York: Pantheon, 2011).

224 *one of the nation's most popular outdoor activites*: United States Fish and Wildlife Service, "National Survey of Fishing, Hunting, and Wildlife-Associated Recreation: National Overview" (2012), http://digitalmedia.fws.gov/cdm/ref/collection/document/id/858.

224 *Worldwide, more than one in ten humans*: Stephen J. Cooke and Ian G. Cowx, "The Role of Recreational Fishing in Global Fish Crises," *BioScience* 54 (2004): 857–59.

225 *recreational fishing is big business*: American Sportfishing Association, "Recreational Fishing: An Economic Powerhouse" (2013), http://asafishing.org/facts-figures.

225 *Eye damage from hooks*: Robert B. DuBois and Richard R. Dubielzig, "Effect of Hook Type on Mortality, Trauma, and Capture Efficiency of Wild, Stream-Resident Trout Caught by Angling with Spinners," *North American Journal of Fisheries Management* 24 no. 2 (2004), 609–16; Robert B. DuBois and Kurt E. Kuklinski, "Effect of Hook Type on Mortality, Trauma, and Capture Efficiency of Wild, Stream-Resident Trout Caught by Active Baitfishing," *North American Journal of Fisheries Management* 24, no. 2 (2004): 617–23.

226 *Landing nets cause injuries*: B. L. Barthel et al., "Effects of Landing Net Mesh Type on Injury and Mortality in a Freshwater Recreational Fishery," *Fisheries Research* 63, no. 2 (2003): 275–82.

226 *died before they were weighed*: Thomas M. Steeger et al., "Bacterial Diseases and Mortality of Angler-Caught Largemouth Bass Released After Tournaments on Walter F. George Reservoir, Alabama/Georgia," *North American Journal of Fisheries Management* 14, no. 2 (1994): 435–41.

226 *Nevertheless, a fish usually survives*: "Bring That Rockfish Down," Sea Grant catch-and-release brochure on preventing and relieving barotrauma to fishes, www.westcoast.fisheries.noaa.gov/publications/fishery_management/recreational_fishing/rec_fish_wcr/bring_that_rockfish_down.pdf.

226 *reduced the biomass of predatory fishes*: David Shiffman, "Predatory Fish Have Declined by Two Thirds in the Twentieth Century," *Scientific American*, October 20, 2014, www.scientificamerican.com/article/predatory-fish-have-declined-by-two-thirds-in-the-20th-century.

227 *Sylvia Earle puts it this way*: Evans, "Making Waves."

227 *A tuna consumes her body weight*: Valérie Allain, "What Do Tuna Eat? A Tuna Diet Study," SPC Fisheries Newsletter 112 (January/March 2005): 20–22.

227 *Atlantic and Pacific bluefin tunas*: Ira Seligman and Alex Paulenoff, "Saving the Bluefin Tuna" (2014), https://prezi.com/lhvzz56yni7_/saving-the-bluefin-tuna.

227 *that's twice the price of silver*: British Broadcasting Corporation (BBC), "Superfish: Bluefin Tuna" (2012), a forty-four-minute documentary, can be viewed at: http://wn.com/superfish_bluefin_tuna.

227 *notably pregnant and nursing women*: FAO Fisheries and Aquaculture Department, "Fish Contamination," accessed October 9, 2015, at www.fao.org/fishery/topic/14815/en.
228 *Among the undesirable effects*: "Fish," NutritionFacts.org, accessed October 2015, http://nutritionfacts.org/topics/fish.
228 *people in developed countries*: David J. A. Jenkins et al., "Are Dietary Recommendations for the Use of Fish Oils Sustainable?" *Canadian Medical Association Journal* 180, no. 6 (2009): 633–37.
228 *The main problem with this advice*: Jenkins et al.
228 *increase imports from developing countries*: Jenkins et al.
228 *Having witnessed the sharp decline*: Natasha Scripture, "Should You Stop Eating Fish?" IDEAS.TED.COM, August 20, 2014, http://ideas.ted.com/should-you-stop-eating-fish-2.
228 *"Ask yourself this," she says*: Sylvia Earle, in Scripture, "Should You Stop Eating Fish?"
228 *Populations of some commercially heavily exploited*: Alister Doyle, "Ocean Fish Numbers Cut in Half Since 1970," *Scientific American*, September 16, 2015, www.scientificamerican.com/article/ocean-fish-numbers-cut-in-half-since-1970/?WT.mc_id=SA_EVO_20150921.
229 *"If an animal is sentient"*: Vonne Lund et al., "Expanding the Moral Circle: Farmed Fish as Objects of Moral Concern," *Diseases of Aquatic Organisms* 75 (2007): 109–18.

EPILOGUE

231 *"The arc of the moral universe"*: Martin Luther King, "Keep Moving from This Mountain," sermon at Temple Israel (Hollywood, CA, February 25, 1965). Taken from https://en.wikiquote.org/wiki/Martin_Luther_King,_Jr.#Keep_Moving_From_This_Mountain_.281965.29.
232 *"Fish are always in another element"*: Foer, *Eating Animals* (New York: Back Bay Books, 2010).
234 *Despite what we find in headlines*: Steven Pinker, *The Better Angels of Our Nature: Why Violence Has Declined* (New York: Viking Penguin, 2011).
234 *Local ordinances have changed animals'*: www.coloradodaily.com/ci_13116998?source=most_viewed. The "guardian campaign" website was last updated in 2012: www.guardiancampaign.org; www.guardiancampaign.org/guardiancity.html.
234 *judge held a hearing*: David Grimm, "Updated: Judge's Ruling Grants Legal Right to Research Chimps," last updated April 22, 2015, http://news.sciencemag.org/plants-animals/2015/04/judge-s-ruling-grants-legal-right-research-chimps. The judge later reversed her decision. Jason Gershman, "Judge Says Chimps May One Day Win Human Rights, but Not Now," July 30, 2015, http://blogs.wsj.com/law/2015/07/30/judge-says-chimps-may-one-day-win-human-rights-but-not-now.

234 *In parts of Europe it is now unlawful*: The northern Italian town of Monza enacted such a law in 2004, www.washingtonpost.com/wp-dyn/articles/A44117-2004Aug5.html. Rome followed suit in 2005, www.cbc.ca/news/world/rome-bans-cruel-goldfish-bowls-1.556045.

234 *A law enacted in April 2008*: Accessed November 2015 at: www.swissinfo.ch/eng/life-looks-up-for-swiss-animals/6608378; www.animalliberationfront.com/ALFront/Actions-Switzerland/NewLaw2008.htm.

234 *In Germany, a 2013 law*: Anonymous (2012). *Tierschutz-Schlachtverordnung*, vom 20 (December 2012): BGBl. I S. 2982.

235 *In Norway, the use of carbon dioxide stunning*: FishCount.org, "Slaughter of Farmed Fish," http://fishcount.org.uk/farmed-fish-welfare/farmed-fish-slaughter, accessed December 11, 2015.

236 *"When I see a salmon farm"*: Paul Watson, personal communication, May 2015.

Acknowledgments

To Amanda Moon at FSG, for your vision, your advocacy, your guidance, and your constant support. Your excitement and positive energy throughout the project have meant more than you could know.

To Stacey Glick, for seeing the potential of this project, and for taking care of your clients at least as well as cleanerfishes take care of theirs.

To Annie Gottlieb, for an incredibly thorough and incisive copyedit.

To my reviewers, Ken Shapiro, Martin Stephens, Jeannie Geneczko, Reggie Adams, Culum Brown, Marilyn Balcombe, Peter Hagen, Karen Diane Knowles, and Tierney Thys, for your wisdom and your constructive comments.

To Scott Borchert, Stephen Weil, and Laird Gallagher at FSG, for your professionalism and for always being ready to assist.

To Culum Brown, Bernd Kramer, Gordon Burghardt, Ted Pietsch, Nafsika Karakatsouli, Sharon Young, Chris Good, Cristina Zenato, Alan Goldberg, Ron Thresher, Iris Ho, Gordon O'Brien, K. K. Sheenaja, Roman Kolar, Erin Williams, Jay Stauffer, Victoria Braithwaite, Billo Heinzpeter Studer, Lynne Sneddon, Tierney Thys, Rene Umberger, Lynton Burger, Ila France Porcher, Scott Gardner,

Stephanie Cottee, Bill Loftus, Dos Winkel, Joe Denham, Captain Paul Watson, Stephen Corbett, Robert Wintner, Yvonne Sadovy, Marian Wong, Joan Dunayer, Robert Warner, Michael Engel, and John Lucas, for sharing your expertise.

To Rae Sikora, Sabrina Golmassian, Alexandra Reichle, Teresa Fisher, Sarah Kindrick, Karen Day, Robin Walker, Karen Cheng, Ben Callison, Ila France Porcher, Jamie Cohen, Holly Fernandez Lynch, Ana Negrón, Neville Jacobs, John Peters, Heleanna Amicone, Mike Howell, Lori Cook, Lori Williamson, Rosamonde Cook, Methea Sapp, Mo Dawley, Vicky Thornley, Ingrid Newkirk, Cathy Unruh, Tali Ovadia, and Dave Bonnell, for sharing your stories.

To Katherine Head, for organizing the citations.

To Maureen Balcombe, Reggie and Marlie Adams, Andrew Rowan, Lori Marino, Anthea and Joe Messersi, Marilyn and Emily Balcombe, Cindi Lostritto, Sonia Faruqi, Laura Moretti, Marc Bekoff, Melanie Joy, Sabrina Brando, and Bruce Friedrich, for the myriad ways you provided support and inspiration.

To the fishes, for adding beauty and mystery to the world.

Index

Page numbers in *italics* refer to illustrations.